Cocoa Production and Processing Technology

Cocoa Production and Processing Technology

Edited by **Raj Patil**

Printed in 2017

ISBN: 978-1-68117-029-9

Library of Congress Control Number: 2015931396

SCITUS Academics LLC,
616, Corporate Way, Suite 2, 4766,
Valley Cottage, NY 10989

www.scitusacademics.com

Contents

Preface

This book presents innovative techniques for sustainability and traceability in high-quality cocoa production and explores new product development with potential for cost reduction as well as improved cocoa bean and chocolate product quality. Cocoa trees resemble English apple trees. They grow best under the canopy of tropical rainforests, seldom reaching more than 7.5 metres (25 feet) high. To flourish they need to be shaded from direct sun and wind, particularly in the early growth stages. The cocoa tree has broad, dark leaves about 25cm long, and pale-coloured flowers from which bean pods grow. A native of the central and South American rainforests, cocoa trees are now cultivated in many tropical locations around the world. Two methods are generally used to establish cocoa tree plantations. Cocoa trees begin to bear fruit when they are three to four years old. They produce pink and white flowers throughout the year, growing in abundance after before the rain starts. However the pods grow straight out of the trunk and the main branches, which is most unusual. Only a small proportion of the flowers develop into fruit over a period of about five months. The trees are carefully pruned so that pods can be more easily harvested. The harvesting of cocoa pods is very labour-intensive. On West African small-holdings the whole family, together with friends and neighbours help out. Ripe pods are gathered every few weeks during the peak season. The high pods are cut with large knives attached to poles, taking care not to damage nearby flowers or buds. The pods are collected in large baskets, which workers carry on their heads, and piled up ready for splitting. The book examines the fairtrade and organic cocoa industries and their influence on smallholder farmers. The chapters provide in-depth coverage of cocoa cultivation, harvesting and post-harvest treatments with a focus on cocoa bean composition, genotypic variations and their influence on quality, post-harvest pre-treatments, fermentation techniques, drying, storage and transportation.

Editor

Chapter 1

Impact of Cocoa Processing Technologies in Free Fatty Acids Formation in Stored Raw Cocoa Beans

Simplice Tagro Guehi[1], Michael Dingkuhn[2], Emile Cros[3], Gérard Fourny[3], Robert Ratomahenina[4], Guy Moulin[4] and Anne Clement Vidal[2]

[1]Laboratoire de Biochimie et de Technologie Alimentaire. UFR Sciences et Technologies des Aliments. Université d'Abobo-Adjamé. 02 Bp 801 Abidjan 02. Côte d'Ivoire.

[2]CIRAD-AMIS, Equipe Ecotrop, TA 40/01, 34398 Montpellier Cedex 5, France.

[3]CIRAD-CP/Programme Cacao, Laboratoire de Chimie-Technologie du cacao, TA 80/16, 34398 Montpellier Cedex 5,France.

[4]INRA-Montpellier, Laboratoire de microbiologie industrielle et de génétique des microorganismes, 2 Place Viala, 34000 Montpellier, France.

ABSTRACT

The quality of raw cocoa beans depends widely on their free fatty acids (FFA) content. High FFA content is a serious quality defect and reduces the technical and economic value of the cocoa beans. The work investigates the influence of cocoa processing technologies on FFA formation during storage of raw cocoa beans. Different samples of ferment dried cocoa beans purchased from Cote d'Ivoire were stored and analysed for FFA content. Very low FFA contents were found in whole healthy cocoa beans generally complied with UE standards (1.75% oleic acid equivalent) throughout storage while high FFA content was found in poor quality and broken healthy beans. The formation of FFA did not depend on the genotype or on cocoa post-harvest processing technologies. However, high and increasing FFA contents were observed in defective cocoa beans and could be attributed probably to the activity of microflora which in turn were associated with initial quality and loss of physical integrity of the cocoa beans.

INTRODUCTION

Cocoa beans are actually seed from the fruit of Theobroma cacao tree. Cocoa butter, the naturally fat, contained in cocoa beans is the only continuous phase of chocolate. It is responsible for dispersion of the other constituents of chocolate properties during storage, handling and tasting (Nickless, 1996). Cocoa butter was currently reported to be the main vegetable fat used in the chocolate manufactures because of its rheological, textural and chemical characteristics such as triglycerides fatty acids composition, (Awua, 2002; Lipp and Anklam, 1998; Whithefield, 2005). Its hardness depends on the saturated and unsaturated fatty acid contents bound in triglycerides, and on free fatty acids (FFA) content. The general opinion is that higher FFA content leads to a decrease in hardness of cocoa butter (Pontillon, 1998) and must be considered as a raw cocoa commercial value reducing factor both for producers and chocolate manufacturers. For reasons of quality therefore, the directive 73/241/EEC (EEC, 1973) limits the maximum FFA content to 1.75% oleic acid equivalent in cocoa butter. Yet, for several years, from 15 to 20% of annual Ivorian cocoa production is considered to have excessive FFA contents (exporters' statistics) recurrently and seasonally, notably at the end of the main season. FFA are carboxylic acids released from triglycerides (Selamatet al., 1996) through the effect of a lipase (E.C. 3.1.1.3) or an oxidation.

The risks of oxidation are negligible in cocoa butter due to its low unsaturated fatty acid content (Withefiled, 2005) and high content of polyphenols, natural antioxidants, in cocoa beans (Nickless,1996). To date, although no lipase activity has yet been detected in fermented and dried cocoa beans to our knowledge, the role of a lipase is strongly suspected. Numerous studies have shown plant lipase activity is mostly detectable during seed germination (Wanasundara et al., 2001), except in some oil crops such as the castor-oil plant (Ory et al., 1962), nigella kernel (Mert et al., 1995) and rice (Raghavendra and Prakash, 2002). For Fowler (1999), high FFA contents in cocoa beans might result from black beans originating from rotten pods or germinated beans. Likewise, microflora, particularly moulds, can cause similar problems during storage (Hiol, 1999). Indeed, Wood and Lass (1985) and Pontillon (1998) suggested that FFA occurrence in stored cocoa beans is linked to the action of microbial lipases. Therefore, this study aims to determine the impact of cocoa processing technologies on FFA formation in raw cocoa beans and if microflora of the cocoa beans are implied in FFA formation.

MATERIALS AND PROCESSING

Cocoa beans used in this study were originated healthy from region of Soubré in Côte d'Ivoire. A total of 16 samples (25 pods each) were opened and beans of each protocol were fermented using the micro-fermentation technique in a wooden box measuring 65 x 50 x 50 cm (turned manually after 48 and 96 h). Cocoa beans were sun dried on plastic tarpaulins for one week, reducing bean moisture content from 60 to 8%. All samples were stored in a climatic cabinet (Firlabo, SB-BVEHF type). Temperature and relative humidity were fixed respectively at 27°C and 75% for 6 months. For storage, 50 g of cocoa beans were individually putted in bags to be removed at each sampling time, 0, 2, 3, 4 and 5 months later. FFA content was measured on carefully shelled, manually and finely ground beans (< 500 μm). Effect of cocoa beans processing technologies on FFA's formation was studied. Genetic origin's effect was studied using beans from three cultivars ripe pods:

Amelonado, Ivorian 1 st generation of hybrids (Amelonado x West African Trinitario) and open pollinated progenies. Influence of pod degree ripeness was studied using beans from pre-ripe, ripe and over-ripe pods of Ivorian 1st generation of hybrids. Pods were opened 5 days after harvesting. Beans from 1st generation hybrids pods harvested the same day were used for the impact of the time lapse between pods harvesting and opening. Pods were divided into three parts and were opened respectively the day of harvesting, 5 and 9 days after harvesting.

Ripe pods from open pollinated progenies were used for study of fermentation duration's effect. Beans samples were taken in the middle of the cocoa beans mass each day throughout fermentation period and then were sun dried. Effect of raw cocoa beans quality on FFA formation was studied using 1200 g of uncontrolled cocoa samples of three different quailties: healthy beans which are brown and compact with no visible defects, clustered beans and black beans originated from Abidjan harbour. Clustered beans are the agglomerates of 2 to several beans. Each sample was divided into 4 fractions: beans of first 2 fractions were individually and roughly broken and those of second 2 fraction remained whole. And then both sub-samples were stored (27°C, RH 75%). FFA content was measured on 50 g sampled from each fraction every two weeks. About 1800 g of black beans samples were used in order to determine influence of decontamination of the cocoa beans and beans physical integrity in the FFA's formation in raw cocoa beans. The samples were divided into 6 equal portions: 2 untreated cocoa beans of first fraction were stored as previously described, the beans of the 2 others parts are individually and roughly broken (control) and cocoa beans of last 2 fractions were broken and then individually and carefully decontaminated by soaking in a 5% (v/w) hydrogen peroxide solution (Sigma) for 3 hours at 28°C. After sterilely drying at 27°C during 3 days about 40 g of each decontaminated cocoa beans were individually sealed hermetically into sterile flasks prior to storage for 12 weeks as previously described. FFA content was measured every two weeks on samples from one flask per experimental protocol and from over corresponding subsamples.

Physicochemical Analyses

About 15 g of cocoa beans were carefully shelled manually. Cocoa nibs were frozen in liquid nitrogen before finely grinding in a kitchen- scale coffee grinding (Moulinex, France) to the smallest particle (size < 500 µm). Ten grams of cocoa powder were putted in Whatman cartridge and soaked in 350 ml of petroleum ether (Prolabo Normapur, type 40-60°C) for one night. Cocoa butter was extracted on a Soxhlet apparatus for 8 h. After eliminating of solvent in a rotary evaporator, FFA contents were quantified by the official method 42-1993 (IOCCC, 1996). About 5 g of extracted cocoa butter was weighed (W) and dissolved in 50 ml of a previously hot petroleum ether/absolute ethanol mixture (1:1, v/v) neutralized by adding phenolphthalein. Titration carried out with 0.1N alcoholic KOH solution used (V) was noted. FFA (% oleic acid) was calculated as follows: ( )

$$FFA(\%\ oleic\ acid) = \frac{(28.2 \times V \times N)}{W}$$

Statistical Analysis

Data represent the means of triplicate determinations. Analysis of variance were conducted using JMP software Version 5 (SAS Institute, 2002) with mean separation performed by the LSD test ($P<0.05$). Quadratic regression function at 0.05 confidence intervals was conducted using Sigma Plot 8.0 software.

RESULTS AND DISCUSSION

Effect of Cocoa Processing Technologies on the FFA Content

FFA content in whole healthy beans was low and ranged from 0.2 to 0.8%, irrespective of the genotype (Figure 1). FFA contents are not significantly different whatever the genotype of cocoa. In the similar FFA content of pods ripeness degree (results not shown) or pods opening delay are around the same values (Figure 2). Slighter increasing FFA depending on the fermentation duration indicates a difference (LSD test, $P<0.05$) within fermented cocoa beans. Initial

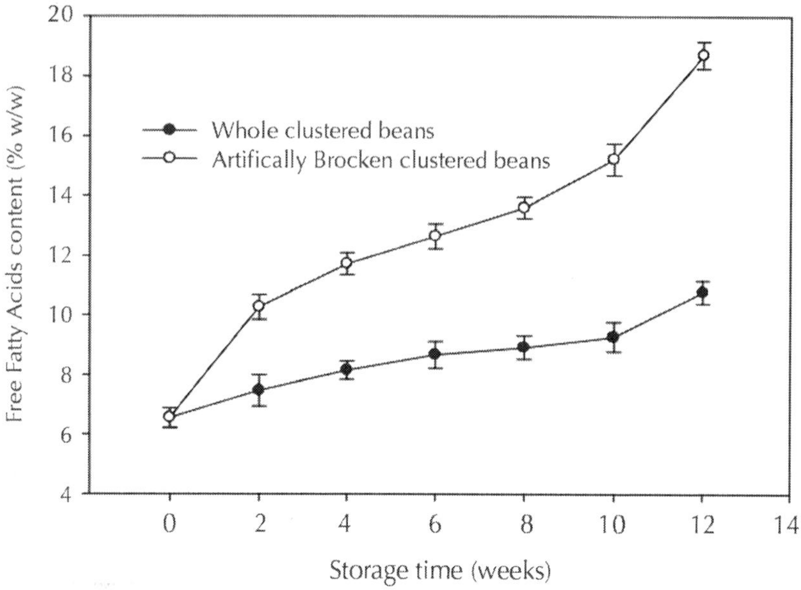

Figure 4: Influence of artificially broken clustered beans on FFA content during storage at

27°C, 75% HR for up to 2 months. Bars represent the standard error of the mean (n=3).

The low FFA content in whole healthy beans increased from 0.48 to 0.78% and did not exhibit an appreciable change during over 12 weeks' storage (results not shown). Considerable increasing FFA contents were observed in poor quality cocoa beans such as clustered beans where it increased from 6.23% to ca 11% (Figure 4). No variation of FFA content was noted in black beans where FFA content remained around 10% (results not shown). Bean clustering could have been caused either by harvesting of unripe pods, or by poor separation of the beans and placentas during cocoa pods opening. So beans would be clustered in placentas during and after fermentation. Barel (1998) have mentioned that following inadequate drying due to a surface crust clustered cocoa beans show always high moisture content which favoured moulds growth.

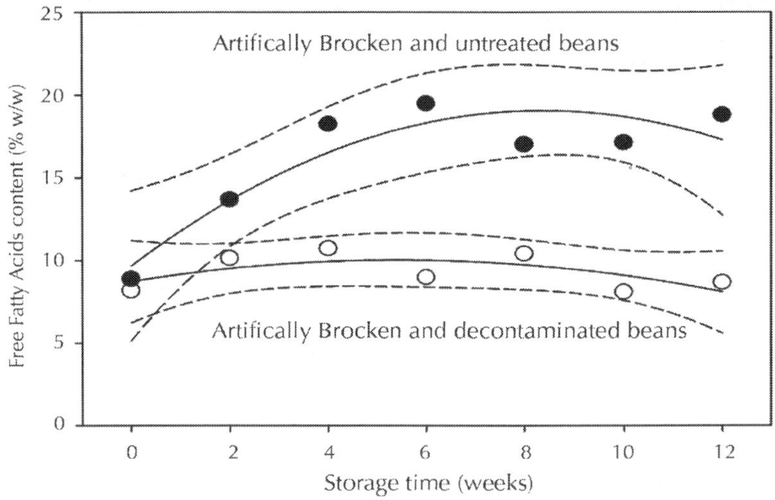

Figure 5: Variation of FFA content in decontaminated artificially broken black beans during storage. Solid lines indicate quadratic regression functions and broken lines indicate 0.05 confidence intervals for each function.

According to Poisson and Cahagnier (1979) and Raghavendra and Prakash (2002) foodstuffs with high moisture content were easily attacked by moulds. These moulds could produce lipase (Wood and Lass, 1985) which in contact with cocoa butter of broken cocoa nibs released FFA from triglycerides. Black beans were probably originating from overripe pods infected by Phytophthora sp. Fungal diseases could create optimal conditions for the development of moulds (Renaud, 1954), which in turn would lead to FFA formation. However, variable FFA content of black bean sample was probably due high variation among individual beans and insufficient sample size. Given the variation of FFA content due probably to a highly heterogeneity of samples, it was not possible to characterize changes in FFA contents in storage conditions used. High FFA contents were due to poor quality of cocoa raw materials quality as previously demonstrated (Guénot et al., 1976). Slighter increasing of lower FFA contents in whole healthy beans than that in defective beans confirms that storage time did not affect FFA formation if processing technologies were previously appropriate. The greater the initial FFA content of the raw cocoa beans, or the lower the quality of the beans, the greater is this increase in FFA. And when cocoa beans were artificially broken, their FFA content increased strongly whatever cocoa beans initial quality. FFA content of decontaminated artificially broken beans remained unchanged and well below the considerably increasing FFA content in the untreated cocoa beans. Using the quadratic regression functions at 0.05 confidence intervals indicates that the variations over storage time in FFA content of untreated cocoa beans sample and decontaminated sample are significantly different (Figure 5).

As decontamination prior to storage under sterile conditions stopped FFA formation in broken black beans only lipase activity of external origin can thus be considered. This conclusion was supported by many studies which demonstrated that plants in which lipases are detected during dormancy are rare. This is the case with the castor-oil plant (Ory et al., 1962) and nigella (Mert et al., 1995). Indeed, enzyme lipase is responsible for lipid breakdown in plants to produce the energy required for embryo growth (Imeson et al., 1993). Yet, during cocoa beans fermentation, the rise in temperature (≥50°C) and diffusion of acid products through the cotyledons due to partial cell walls lyses, hence death of germ and prevent germination (Barel, 1998).

In addition fermentation process induced contact between proteins and polyphenolic compounds hence inactivation of enzymatic proteins. Indeed many polyphenolic compounds in plants were already described as enzyme activities inhibition factor (Hagermann, 1992; Kallithraka et al., 2001; Misnawi et al., 2002). So cocoa beans endogenous enzymes could not be active in complete ferment dried cocoa beans (Yusep et al., 2002). Cocoa beans of poor quality deteriorate much faster than cocoa beans of hight quality. Circumstantial evidence was presented that lipolytic activity leading to the FFA formation was due to microbes associated with other factors such as the quality, the physical condition, and the storage conditions of cocoa beans (Guénot et al., 1976, Oyeniran, 1979; 1980) and not to endogenous plant lipases by decontaminating part of the cocoa beans with hydrogen peroxide. Similar results were obtained by Hiol (1999) in previous study on formation of FFA in palm fruits caused by moulds. Thus cultivation and identification of microbes responsible of FFA formation in raw cocoa beans was not elucidated in this study, yet the relatively high FFA content exhibited by the poor quality beans and no change observed in FFA content of decontaminated beans were good evidence to support their potential lipase-producing abilities and their action on cocoa butter.

ACKNOWLEDGEMENTS

This research was supported by the Cirad cocoa program Montpellier. We are grateful to Touton Company, Industrial private cocoa producer for providing cocoa pods and ADM cocoa for their logistical support and providing of uncontrolled cocoa samples.

REFERENCES

1. Awua PK (2002). Cocoa processing and chocolate manufacture in Ghana. Essex. David Jamieson and Associates Press Inc.
2. Barel MA (1998). Première transformation du cacao. Formation del'arôme cacao. In J. Pontillon. Cacao, chocolat, production, utilisation, caractéristiques. (pp. 96-115). Paris: Techniques et documentation,Lavoisier.
3. EEC (1973). Directive 73/241/EEC by European Parliament and the European Council relating to cocoa and chocolate products intended for human consumption. Official Journal of the European Communities L 228 of 16/08/1973, pp. 0023-0035.
4. Fowler MS (1999). Cocoa beans: from tree to factory. In: Beckett ST (Ed). Industrial chocolate manufacture and use, Oxford: Blackwell Science, pp. 8-35.
5. Guénot MC, Perriot JJ, Vincent JC (1976). Evolution de la microflore et des acides gras des fèves de cacao au cours du stockage: étude préliminaire. Café, Cacao, 20: 53-58.
6. Hagermann AE (1992). Tannin-protein interaction. In Ho CT, Lee CY,Huang MT (Eds). Symposium on Phenolic compounds in food and their effect on health: Analysis, occurrence and chemistry. ACS Series. 506: 237-247.
7. Hiol A (1999). Contribution à l'étude de deux lipases extracellulaires issues de souches fongiques isolées à partir du fruit de palme PhD Dissertation, Université d'Aix, Marseille, France.
8. Imeson HC, Rood SB, Weselake RJ, Zanewichn KP, Bullock BA, Stobbs KE Kocsis MG (1993). Hormonal control of lipase activity in oilseed rape germinant. Physiol. Plant. 89: 476-482. Guehi et

al. 179

9. IOCCC. International Office of Cocoa, Chocolate and Sugar Confection. (1996). Determination of free fatty acids (FFA) content of cocoa fat as a measure of cocoa nib acidity. Analytical Method n°42-1993.
10. Kallithraka S, Bakker J, Cliffor MN (2001). Interaction of (+)-catechin, (-)-epicatechin, procyanidin B2 and procyanidin C1 with pooled human saliva in vitro. J. Sci. Food Agric. 81: 261-268.
11. Lipp M, Anklam E (1998). Review of cocoa butter and alternative fats for use in chocolate- Part A. Compositional data. Food Chem, 62: 73-79.
12. Mert S, Dandik L, Aksoy HA (1995). Production of glycerides from glycerol and fatty acids by native lipase of Nigella sativa seed. Appl.Biochem. Biotechnol. 50: 333-342.
13. Misnawi JS, Nazamid S, Jamilah B (2002). Activation of remaining key enzymes in dried under-fermented cocoa beans and its effect on aroma precursor formation. Food Chem. 78: 407-417.
14. Nickless H (1996). Cocoa butter quality. In: Selamat J, Lian BC, Lai TK, Ishak WRW, Mansor M (eds). Proceeding of the Malaysian international cocoa conference Kuala, Lumpur. pp. 322-336.
15. Ory RL, Saint-Ange AJ, Altschu AM (1962). The acid lipase of castor, properties and substrat specificity. J. Lipids Resour. 3: 99-105.
16. Oyeniran JO (1980). The role of fungi in the deterioration of tropical stored products. Nigerian stored products research Institute (Lagos), Occasional paper series n°2: 9-10.
17. Oyeniran JO (1979). The influence of the storage environment on the quality of commercial cocoa with special reference to mouldiness of the beans. Proceedings of 7th International Conference on cocoa research Douala, pp. 577-582.
18. Poisson J, Cahagnier B (1979). Développement des moisissures sur les graines en fonction de diverses caractéristiques du substrat et de la population fongique. Tech. Indust. Céréalières, 169: 3-8.
19. Pontillon J (1998). Le beurre de cacao et les matières grasses en chocolaterie. In : Pontillon J (ed) Cacao, chocolat, production, utilisation, caractéristiques (p. 326). Techniques et documentation,

Lavoisier, Paris, pp. 257-269.

20. Raghavendra MP, Prakash V (2002). Phenylcolboronic acid-a potent inhibitor of lipase from Oryzae sativa. J. Agric. Food Chem. 50: 6037-6041.
21. Renaud R (1954). La qualité du cacao. Les moisissures des fèves fermentées. Agronomie tropicale (Nogent-sur-Marne), 9 : 563-583. SAS Institute, Inc. (2002). JMP Software Version 5, Cary, NC.
22. Selamat J, Hamid MA, Mohamed S, Man, CY (1996). Physical and chemical characteristics of Malaysian cocoa butter. In Selamat J,Lian BC, Lai TK, Ishak WRW, Mansor M (eds). Proceeding of the Malaysian international cocoa conference Kuala Lumpur. pp. 351-357.
23. Wanasundara PD, Wanasundara UN, Shahidi F (2001). Lipolytic activity of enzymes from germinating seeds of sesame (Sesamum incidcum L.). J. Food Lipids. 8: 75-84.
24. Whitefield R (2005). Making chocolates in the factory. Kenedy's Publications Ldt, London.
25. Wood GA, Lass RA (1985). Cocoa. Logman Scientific and Technical,Essex.
26. Yusep I, Jinap S, Jamilah B, Nazamid S (2002). Influence of carboxypeptidases on tree amino acid, peptide and methylpyrazine contents of under-fermented cocoa beans. J. Sci. Food Agric. 82: 1584-1592.

Chapter 2

Applying Life Cycle Management of Colombian Cocoa Production

Oscar Orlando Ortiz-R[I],*; Raquel Amanda Villamizar Gallardo[II]; Joshua Mauricio Rangel[III]

[I]Department of Industrial Engineering, University of Pamplona, Pamplona, Colombia

[II]Department of Microbiology, University of Pamplona, Pamplona, Colombia

[III]Group of Environmental Management Analysis, University of Pamplona, Pamplona, Colombia

ABSTRACT

The present research aims to evaluate the usefulness of the application of Life Cycle Management in the agricultural sector focusing on the environmental and socio-economic aspects of decision making in the Colombian cocoa production. Such appraisal is based on the application of two methodological tools: Life Cycle Assessment, which considers environmental impacts throughout the life cycle of the cocoa production system, and Taguchi Loss Function, which measures the economic impact of a process' deviation from production targets. Results show that appropriate improvements in farming practices and supply consumption can enhance decision-making in the agricultural cocoa sector towards sustainability. In terms of agri-business purposes, such qualitative shift allows not only meeting consumer demands for environmentally friendly products, but also increasing the productivity and competitiveness of cocoa production, all of which has helped Life Cycle Management gain global acceptance. Since farmers have an important role in improving social and economic indicators at the national level, more attention should be paid to the upgrading of their cropping practices. Finally, one fundamental aspect of national cocoa production is the institutional and governmental support available for farmers in face of socio-economic or technological needs.

INTRODUCTION

The objectives of concepts intended to asses and implement sustainability in agriculture are to consolidate the complex and diverse principles of this theoretical paradigm and to transform them into recommendations for agricultural practices (Von Wirén-Lehr, 2001). Regarding agro-food industrial activities, research around the world has introduced different tools for measuring the impact of production processes on the framework provided by product life cycle analysis. These research initiatives have proposed improvements in all production stages in order to boost environmental performance as a whole. Such improvements not only seek to meet consumer demands for environmentally friendly products, but also to increase productivity and competitiveness of green product markets (Mattsson et al., 2000; Huijbregts et al., 2001; Heller & Keoleian, 2003; Black et al., 2011;

Carof et al., 2013). Cocoa (*Theobroma cacao* L) is a very popular fruit due to the fact that all kinds of chocolates and confectionaries are made from its beans (Efraim et al., 2010). It is a tropical crop originated from Central and South America grown under humidity conditions that produces from 0.5 to 2 kg of dried and fermented cocoa beans per tree annually (Medeiros & Lannes, 2010).

Currently, Colombia is the fifth producer worldwide and the third one in Latin America. According to the Cocoa Development Ten Year Plan 2012-2021, in Colombia, there are 660,000 hectares available for growing this crop. Requiring an investment of $2.5 billion (Colombian peso) in the next ten years, these lands production is projected to reach 246,000 tons by the year 2021. Following studies by the Ministry of Agriculture, the sector currently employs 70,000 workers and is projected to recruit 40,000 more by 2014, 130,000 at the end of the plan (2021), and 100,000 in the medium term (Federación Nacional de Cacaoteros, 2012).

Nevertheless, in Colombia there is need for research on the productivity of this crop, whose production system has not been modernized. According to FAOSTAT, in 2011, the cocoa cultivation area in Colombia was 99,205 hectares, with a productivity of 446 kg·ha, which is certainly a small figure when compared to that of the first producer worldwide (Ivory Coast: 700 kg·ha) (Food and Agriculture Organization of the United Nations, 2011). This low productivity, which results from the old hybrid materials grown in most Colombian plantations, has consequently determined the low international competitiveness featuring this product in Colombia. Contrasts are also outstanding at the national level. In 2010, the department of "Norte de Santander (N. de S)" reached an average yield of 477 $kg \cdot ha^{-1}$, while the department of Santander reached 564 $kg \cdot ha^{-1}$ in the same year. Furthermore, these are both low figures when compared to the national production objective: 1,500 $kg \cdot ha^{-1}$(Colombia, 2010).

Looking forward to improve sustainability indicators, the agro-food industry is in need of methodologies that are capable of incorporating information on environmental impacts and life cycle costs (Bélanger et al., 2012). Life Cycle Management (LCM) provides an adequate framework for analyzing and managing the sustainability of goods and services. This approach has been used by global businesses to improve the environmental, social, and economic performance of their offerings in order to ensure a more sustainable value chain (United Nations Environment Programme, 2011).

Therefore, the present research aims to evaluate the use of LCM in the agricultural sector focusing on the environmental and socio-economic aspects of decision making in the Colombian cocoa production. Such appraisal is based on the application of two methodological tools: Life Cycle Assessment (LCA) and Taguchi Loss Function (TLF).

LCA is a methodology that follows ISO 14040/44 standards and seeks to provide an assessment of the environmental performance of a product or service throughout its life cycle, from cradle to grave: extraction of raw material, production or processing, use, and subsequent disposal (International standardization Organization, 2006).

Among the recent LCA agricultural studies (Mouron et al., 2006; Zufia & Arana, 2008; Pardo & Zufia, 2012; Roer et al., 2012; Ruviaro et al., 2012), only one study evaluated the application of LCA in the cocoa production in Ghana, in which the entire system was required to produce and process 1 kg of cocoa (Ntiamoah & Afrane, 2008). On the other hand, TLF quantifies the variability of a given process. It has been used to analyze the interaction between the economics of production and process quality in the manufacturing industry (Castillo-Villar et al., 2012). TLF is an effective tool for justifying and evaluating the effects of quality improvements such as tolerance tightening strategies, whose effects on quality have been assessed from an economic standpoint (Berk & Berk, 2000; Bernardo et al., 2001; Ozdemir & Maghsoodloo, 2004; Liao & Kao, 2010). Nevertheless, those studies have not yet been applied to the agro-food industry and limited research on TLF has been published.

In this context, the main objective of the current paper was to develop a proper and comprehensive perspective of cocoa production sustainability by means of Life Cycle Management analysis applied to a case study in Colombia. Therefore, current research was used to draw up LCM-based guidelines integrating social, economic, and environmental indicators of sustainability within the cocoa industry. This study is intended to be a useful LCM reference for diverse stakeholders such as farmers, engineers, and environmentalists as well as government and LCA advisors.

METHODOLOGY

Life Cycle Assessment (LCA) was used to evaluate the environmental impact of the cocoa production system and Taguchi Loss Function (TLF) was applied to measure the economic impact, which is actually a way of estimating benefits. The social implications of the analysis are described next and conclusions are drawn.

Environmental Aspects of LCA

LCA comprises four steps: goal and scope definition, inventory analysis, impact assessment, and interpretation.

Goals and Scope Definition

This step addresses the definition of the system under assessment, the limits of the study (*i.e.*, the processes to be analyzed), and the functional unit.

Therefore, the system comprises the cocoa production process from nursery sowing and includes site preparation and final planting to fertilization, phytosanitary management (including insecticides, herbicides and fungicides), and energy consumption (fuel and transport). The functional unit is based on "1 ha of land planted with cocoa with a projected 25 year life span".

Inventory Analysis

The data used in the present study correspond to 2012, when cocoa planting area was about 116,777 hectares. Research has been conducted on the departments with the largest production records at the national level, namely Santander, with 40.7%; *N. de S*, with 8.3% and Antioquia with 5.3%. Social, economic, and environmental data were obtained from 30 farms representing the current state of cocoa farming in each zone and department, which actually corresponds to conventional cropping and currently reaches yearly averages of 400 to 800 $kg \cdot ha^{-1}$. Thus, large farm size or high production records were not the guiding criteria employed to select the productive units.

Table 1 shows the collected inventory data that correspond to energy and supply inputs and outputs of the system in question. They were classified into two categories: a) a questionnaire has been used in order to obtain primary data directly from the farmers regarding supply (fertilizers, nursery bags, pesticides, etc.) and energy (fuels) consumption records, whose values were adjusted to the functional unit and b) secondary data were taken from the ecoinvent database (Ecoinvent, 2004).

Table 1: Inputs and outputs per hectare per year of cocoa production for the 30 farms

Farmer	yield (kw ha^{-1})	Cal (kg)	Plastic bags (kg)	Gasoil (kg)	Fertilizer (kg)	Herbicide (kg)	Insecticide (kg)
y1	3.00E+02	0.00E+00	1.16E+00	0.00E+00	0	Glyphosate = 5.10E+00	0
y2	200E+02	0.00E+00	1.16E+00	1.46E+01	0	0	0
y3	2.00E+02	0.00E+00	1.27E+00	1.76E+01	N-P-K: 15-15-15 = 2.50E+02	Paraquat=3.75E+00	0
y4	3.00E+02	0.00E+00	1.16E+00	0.00E+00	0	Glyphosate = 1.17E+00	0
y5	4.67E+02	3.30E+02	5.80E-01	9.55E+00	0	Glyphosate = 1.20E+00	0
y6	1.33E+02	0.00E+00	1.16E+00	0.00E+00	N-P-K: 15-15-15 =1.00E+02	Glyphosate = 5.10E+00	0
y7	6.70E+01	5.50E+02	0.00E+00	0.00E+00	N-P-K: 18-18-18 = 1.00E+02	0	0
y8	2.33E+02	5.00E+02	1.27E+00	1.85E+00	N-P-K: 15-15-15 =1.50E+02	Paraquat=0.93E+00	0
y9	2.00E+02	0.00E+00	1.27E+00	0.00E+00	N-P-K: 15-15-15 =1.50E+02	0	0
y10	7.00E+02	1.50E+02	0.00E+00	0.00E+00	N-P-K: 15-15-15 =1 50E+02	0	Chlorpyrifos = 1.23E+00

y11	3.66E+02	0.00E+00	1.16E+00	0.00E+00	Urea and DAP = 2.00E+02	0	7
y12	4.80E+02	2.20E+02	0.00E+00	0.00E+00	urea=5.00E+01	0	0
y13	2.50E+02	5.50E+02	1.27E+00	0.00E+00	Urea. TSP and KCl = 1.00E+02	0	0
y14	2.50E+02	0.00E+00	1.27E+00	0.00E+00	Organic = 1.10E+03	0	0
y15	2.25E+02	5.00E+01	1.16E+00	0.00E+00	N-P-K: 10.30-10 is 5.00E+01	0	0
y16	5.00E+02	0.00E+00	9.57E-01	0.00E+00	Organic = 5.00E+02; plus 15-15-15 =1.00E+02	0	Chlorpyrifos = 0.25E+00
y17	6.00E+02	0.00E+00	7.25E-01	0.00E+00	N-P-K: 15-15-15 =1.00E+02 plus Organic = 1.20E+03	Glyphosate = 1.27E+01	0
y18	3.33E+02	6.66E+02	1.16E+00	0.00E+00	Organic =1.20E+03	Glyphosate = 6.35E+00	0
y19	3.33E+02	0.00E+00	1.16E+00	0.00E+00	N-P-K: 20-15-20(1.50E+02)	0	Chlorpyrifos = 1.27E+00
Y20	5.00E+02	0.00E+00	1.27E+00	0.00E+00	Organic = 1.00E+03	0	Chlorpyrifos = 1.23E+00

Y21	2.00E+02	3.00E+02	1.16E+00	0.00E+00	Organic = 2.00E+02	Glyphosate = 1.25E+00	0
Y22	6.00E+02	5.50E+02	1.27E+00	0.00E+00	Urea. DAP and KCl = 11.00E+02	0	Chlorpyrifos = 1.00E+00
Y23	5.71E+02	3.00E+02	7.25E+02	7.98E+00	N-P-K: 15-15-15 =1.00E+02	Glyphosate = 5.10E+00	0
Y24	3.60E+02	3.13E+02	7.25E-01	1.46E+01	Organic = 1.50E+03	0	0
Y25	3.75E+02	0.00E+00	8.10E-01	9.93E+00	Organic = 1.25E+03	0	0
Y26	1.00E+03	1.00E+02	0.00E+00	0.00E+00	Organic = 3.00E+02	0	0
Y27	1.00E+03	2.00E+02	1.16E+00	5.10E+00	Urea. DAP and KCl =3.00E+02	0	0
Y28	5.00E+02	7.50E+02	1.04E+00	0.00E+00	Organic = 2.00E+03	Glyphosate = 1.42E+01	0
Y29	2.50E+02	0.00E+00	1.16E+00	0.00E+00	N-P-K: 15-15-15 = 2.00E+02	1	0
Y30	2.40E+02	1.10E+02	1.27E+00	0.00E+00	Urea. DAP and KCl = 1.50E+02	0	0

Environmental Impact Assessment

The CML 2 baseline 2000 method (Centre for Environmental Studies, 2001) was used to evaluate the environmental profile of the cocoa production system in question (Leiden University, 2001). Along these lines, Global Warming Potential (GWP, measured in Kg of CO_2 equiv.), calculated over the next 100 years, was taken into account due to the importance of this global phenomenon that affects local conditions (United Nations, 2005).

Interpretation

During this stage, the inventory data collected from the farms were transformed into interpretable environmental impact values adjusted to functional unit (1 ha per year). This allowed evaluating the environmental impact of the production system's supply inputs and outputs. The results of this research were presented in Ecopoints and were developed using a LCA Software called LCA-Manager 1.3 developed by SIMPPLE, Spain.

TLF Economic Assessment of the Cocoa Production System

Taguchi's Loss Function (TLF) was used as the nominal-is-best characteristics when *y* is at the target. The function depends on the magnitude of variation, which is allowed in both directions from the target value, and is formulated in Equation 1:

$$L(y) = k(y - T)^2 \qquad (1)$$

Where T is the nominal value of the productivity target specification; *y* is real field productivity; L is the loss associated to a particular difference between *y* and T; and k is the quality loss coefficient, whose value depends on the cost at any specified limits and on the width of such specification, *e.g.*, T ± Δ, where Δ is the customer›s tolerance to deviation of *y* from the target (Phadke, 1989). Thus, L(y) estimates the cost of failing to meet a given quality standard of the product, resulting in deviation from its target value. The loss can be incurred by the customer as scrap costs; by the producer as inappropriate farming practices; or by society in general, as environmental costs. Accordingly, Taguchi's Loss Function allowed an economic analysis of the farms in terms of income not perceived by the producers due to inadequate farming management, which prevents them from reaching the objective value (T). One scenario was defined through variations in T, the departmental (T_1) productivity goals. The four leading criteria used for direct cost calculation were labor, inputs, equipment and tools, and other costs (Table 2).

Table 2: Leading criteria for direct cost calculation

	Year 1	Year 2	Year 3	Year 4	Year 5	Year 6	Year 7-25	Total
Labor	3.06E+06	2.00E+06	1.69E+06	2.15E+06	2.54E+06	2.82E-4-06	4.08E+07	5.51E+07
Inputs	3.19E+06	8.09E+02	1.03E+06	1.01E-4-06	1.43E+06	1.43E+06	2.02E+07	2.91E+07
Equipment and tools	3.00E+02	9.40E+01	8.70E+01	1.01E+06	1.62E+02	2.37E+02	2.73E+06	4.62E+06
Other costs	3.92E+02	2.53E+02	2.69E+02	3.36E+02	3.70E+02	3.96E+02	5.60E+06	7.62E+06
Total	6.94E+06	3.15E+06	3.08E+06	4.52E+06	4.50E+06	4.88E+06	6.94E+07	9.64E+07

Moreover, the specification criteria used for calculating the marginal revenues of banana, timber, and cocoa clone productivity per hectare and time period (years) are shown in Table 3.

Table 3: Leading criteria employed to calculate marginal revenues for banana, timber and cocoa clones are in CP (Colombian pesos)

	Year 1	**Year 2**	**Year 3**	**Year 4**	**Year 5**	**Year 6**	**Year 7-25**	**Total**
Cocoa beans (kg)			2.02E+06	3.01E+06	3.61E+06	4.52E+06	8.58E+07	9.90E+07
Horn platain (cluster)		6.40E+06	3.20E+06	1.60E+06				1.12E+07
PCC Cedar Wood pink (m^3)							2.19E+07	2.19E+07
Total	0.00E+00	6.40E+06	5.22E+06	4.61E+06	3.61E+06	4.52E+06	1.08E+08	1.32E+08

Social Aspects

Colombian cocoa farming is a subsistence family activity featured by scarce land property (16 ha in average) that generates 8.5 employments per ha (Mantilla et al., 1996). This aspect was assessed using field surveys, which allowed measuring a series of social indicators such as labor, electric power, basic sanitation, sewer availability, and access to water and communications.

RESULTS AND DISCUSSIONS

Environmental Results Using LCA

The results of the current research showed that appropriate improvements in farming practices and supply consumption are likely to enhance decision making in the cocoa growing sector towards sustainability.

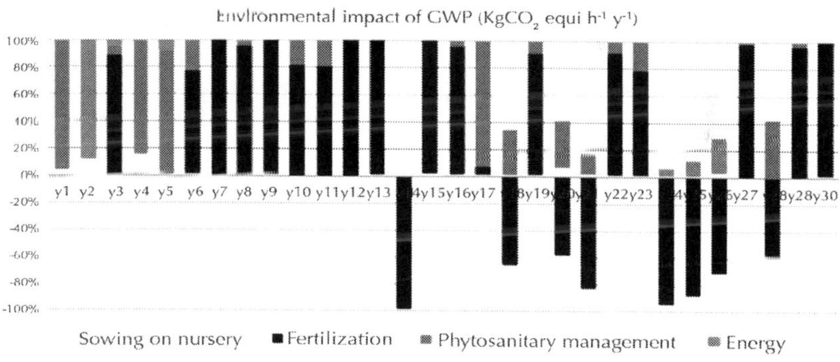

Figure 1: Environmental impact of the studied farms- a GWP analysis.

Figure 1 summarizes each of the studied farmer's environmental profile. The GWP of cocoa production was found to attain positive and negative values. The former correspond to net CO_2 emissions, while negative values represent CO_2 credit. The highest environmental impact resulted from the use of fertilizers, which account for approximately 90-96% of the total life cycle's emissions. The highest emission percentage among the studied farms corresponds to a 250 kg·ha^{-1} dose of a 15-15-15 fertilizer, which far exceeds the records of the other farms employing synthetic fertilizers. Yet, the decision to apply these amendments depends on plantation quality and management, shade intensity, and soil depth and drainage. In some regions, high soil fertility makes it unnecessary to use fertilizers during the first five years of production.

When comparing the use of synthetic and organic fertilization, it can be observed that the latter results in negative GWP values, which implies environmental benefits because the avoided impact that would result from the use of synthetic fertilizer is higher than that of the cropping activity itself. Therefore, there is no doubt that the use of organic fertilizer significantly contributes to reducing environmental impacts such as CO_2 emissions. Yet, some authors have stated that this situation is not really desirable since it does not only prevent obtaining higher yields and profits, but it also affects disease resistance (Suárez & Hernández, 2010).

Particular attention was also paid to inputs aimed at phytosanitary management, whose selection has a strong influence on their environmental burden. Thus, the total emissions of the 30 farms studied was 9.32E+02 kgCO_2-Eq, in which herbicides make up 76%, insecticides 22%, and fungicides 1%. Therefore, even when the contribution of phytosanitary inputs themselves is low compared with those of the whole cocoa production process, choosing them carefully and in appropriate doses can lead to important CO_2 emission reductions in this step.

Economic Results

Table 4 shows the economic performance of the studied farms during the cocoa production lifespan. Negative values of *L(y)* indicate occasional losses determined by deviations from (T_1), which is the target value of the departmental productivity goals.

Table 4: Environmental and economic performance of the 30 studied farms according to Taguchi's Loss Function (TLF); cut off criteria yield > 300

Fanner	T1 (Departmental)	K	L (per ha per 25 years)	kg CO_2, Eq (per ha per year)
Y1	4.77E+02	2.76E+02	-8.63E+06	5.61E+01
Y4	4.77E+02	2.76E+02	-8.63E+06	1.46E+01
Y5	5.64E+02	1.92E+02	-1.82E+06	1.43E+02
Y7	4.77E+02	2.76E+02	-4.63E+07	4.17E+02
YI0	4.77E+02	2.76E+02	1.37E+07	1.14E+02
YI1	5.28E+02	2.76E+02	-9.92E+05	6.07E+02
YI2	4.77E+02	2.76E+02	2.48E+03	1.67E+02
YI6	5.64E+02	1.92E+02	7.85E+05	1.38E+02
Y17	5.64E+02	1.92E+02	2.48E+05	1.46E+02
Y18	4.77E+02	2.76E+02	-5.71E+06	-6.06E+01
Y19	4.77E+02	2.76E+02	-5.71E+06	3.08E+02
Y20	4.77E+02	2.76E+02	1.46E+05	-6.27E+00
Y22	4.77E+02	2.76E+02	4.17E+06	2.02E+02
Y23	5.64E+02	1.92E+02	9.39E+03	2.45E+02
Y24	5.64E+02	1.92E+02	-7.99E+06	-1.47E+02
Y25	5.64E+02	1.92E+02	-6.84E+06	-1.13E+02
Y28	5.64E+02	1.92E+02	7.85E+05	-5.41E+01

The corresponding departmental T values were only overcome by 10 (out of the 30 studied) farms, which are thus shown to have reached higher revenues than those of the other 20, whose values ranged from -5.68E+06 to -4.76E+07. Only two farms reached T values above 800 kg, thus coming close to achieving the national objective. In contrast, farm y6, with an L(*y*) score of -4.76E+07, is the one that loses most of the profit, as its productivity is the least likely to reach objective value of T. This is mostly due to the plantation renewal technique, which consists of grafting on old stocks (colloquially known in Spanish as *"injerto en leño viejo"*).

A common factor determining low productivity in all studied farms resulted from the intense precipitations that affected the country during the 2010/2011 cocoa production year. This, coupled to constant temperatures between 25 and 27 °C, favored the development of pathogen fungi, among which *Moniliophthora roreri*determined approximately 40% productivity losses in the studied farms.

Social Results

It was observed that the cocoa farming families have electrical power and potable water utilities; the latter is usually provided by small local aqueducts. In addition, they are provided with adequate primary and secondary roads, whose functions are restricted due to the condition of tertiary roads and local walking paths.

Children and youngsters contribute to the agronomical labor and attend rural schools. Similarly, women combine housework with occasional farming. It is worth noting that depending on farm size and land administration tenancy system (direct or delegated), it is possible to find more than one family in the same production unit. The typical house of these cocoa farming families is built with low quality materials such as clay brick or wood planks walls; Zinc roofs, and cement or soil floors. Moreover, houses are often used for sun drying of the cocoa beans.

The cultural and entrepreneurial orientation and heterogeneity found among the surveyed farmers showed how some of them take better advantage of the available opportunities to obtain higher prices per kg of cocoa beans. One of such opportunities is the "green product" labels, which, by promoting good agronomic management practices, as well as fauna, flora, soil, and water resource conservation, contributes to achieving higher export prices and providing community and environmental benefits. The government plays an important role in the socioeconomic improvement of cocoa growers. For example, in Santander, a $0.22 U.S. dollar subsidy per kg of cocoa bean sold is granted by the department government. In addition, a $0.67 U.S. dollar subsidy is granted at the national level. Thus, the cocoa growers of Santander feel confident with the support they receive for producing at good quality levels. This explains quite well the fact that the department of Santander is the largest cocoa producer in the country.

CONCLUSIONS

There is no doubt that the use of Life Cycle Management as an assessment tool for the agricultural sector is very important for evaluation of environmental loads, thereby improving sustainability indicators. Climate as well as technological, cultural, and socio-economical differences clearly characterizes the agro industrial management process in any context and in any region. This leads to important differences in the LCA results, and it means that any extrapolation of existing European LCA data for the case of a developing country would imply serious errors. However, its function is always the same: production and consumption of foods for its habitants. Therefore, the potential of this methodology both for the evaluation of the environmental performance of cocoa production and for other agricultural purposes cannot be ignored.

Cropping practices and supply consumption vary from one farm to another. In some of them, both factors are minimum or inexistent, while in others they are considered high if compared to those of other farms analyzed in the same department. However, the application of Taguchi's loss function allows concluding that the adoption of poor agronomic practices or lack in adoption of those practices certainly takes a toll on productivity (T value) and quality, both resulting from poor plant nutrition. Finally, a fundamental aspect of national cocoa production is the institutional and governmental support available. For example, in Santander, cocoa is one of the main pillars of the economy, whose production generates rural development in the region. This makes cocoa farming strong and centered on the sustainable management of the crop and the adequate development of harvest and post-harvest labors. Additionally, some of the surveyed growers see their farms not only as a source of family income, but also as an agribusiness. This can be observed by the fact that synthetic agrochemicals are seen by these farmers as the last fertilizing option, which reflects cultural change towards a sustainable and environmentally friendly agribusiness.

ACKNOWLEDGEMENTS

The authors are grateful to the National Federation of Cocoa Growers - FEDECACAO North of Santander, Colombia and farmer for their cooperation in conducting this study and for the financial support provided by the Colombian Administrative Department of Science, Technology, and Innovation - COLCIENCIAS -, Inter-American Development Bank (IDB), and World Bank (WB) BIRF, (Project Reference 0371-2012).

REFERENCES

1. Bélanger, V., Vanasse, A., Parent, D., Allard, G., & Pellerin, D. (2012). Development of agri-environmental indicators to assess dairy farm sustainability in Quebec, Eastern Canada. *Ecological Indicators, 23*,421-430.
2. Berk, J. & Berk, S. (2000). ANOVA, Taguchi, and Other Design of Experiments Techniques. *Quality Management for the Technology Sector*, 106-123.
3. Bernardo, F. P., Pistikopoulos, E. N., & Saraiva, P. M. (2001). Quality costs and robustness criteria in chemical process design optimization. *Computers & Chemical Engineering, 25*(1),27-40.
4. Black, M. J., Whittaker, C., Hosseini, S. A., Diaz-Chavez, R., Woods, J., & Murphy, R. J. (2011). Life Cycle Assessment and sustainability methodologies for assessing industrial crops, processes and end products.*Industrial Crops and Products, 34*(2),1332-1339.
5. Carof, M., Colomb, B., & Aveline, A. (2013). A guide for choosing the most appropriate method for multi-criteria assessment of agricultural systems according to decision-makers' expectations. *Agricultural Systems, 115*,51-62.
6. Castillo-Villar, K., Smith, N. R., & Simonton, R. (2012). A model for supply chain design considering the cost of quality. *Applied Mathematical Modelling, 36*(12),5920-5935.
7. Centre for Environmental Studies - CML. (2001). *CML 2 baseline method*. University of Leiden.
8. Colombia. Ministry of Agriculture and Rural development - MADR. (2010). *Statistical Yearbook of Agriculture and Fisheries Sector*.
9. Ecoinvent. (2004). *Life Cycle Inventories of Production Systems*. Swiss Centre for Life Cycle Inventories.

10. Efraim, P., Pezoa-García, N. H., Jardim, D. C. P., Nishikawa, A., Haddad, R., & Eberlin, M. N. (2010). Influência da fermentação e secagem de amêndoas de cacau no teor de compostos fenólicos e na aceitação sensorial.*Food Science and Technology, 30*(Supl. 1),142-150.
11. Federación Nacional de Cacaoteros. (2012). *Technical Guide for Cocoa Cultivation*. Bogotá.
12. Food and Agriculture Organization of the United Nations - FAO. (2011). *Faostat.*
13. Heller, M. C., & Keoleian, G. A. (2003). Assessing the sustainability of the US food system: a life cycle perspective. *Agricultural Systems, 76*(3),1007-1041.
14. Huijbregts, M. A. J., Guinee, J. B., & Reijnders, L. (2001). Priority assessment of toxic substances in life cycle assessment. III: Export of potential impact over time and space. *Chemosphere, 44*(1),59-65.
15. International standardization Organization - ISO. (2006). *ISO 14040: Environmental management -Life cycle assessment - Principles and framework*. ISO.
16. Leiden University. Institute of Environmental Science. (2001). *CML's impact assessment methods and characterization factors*. Centre for Environmental Studies, University of Leiden, CML 2 baseline method.
17. Liao, C. N., & Kao, H. P. (2010). Supplier selection model using Taguchi loss function, analytical hierarchy process and multi-choice goal programming. *Computers & Industrial Engineering, 58*(4),571-577.
18. Mantilla, J., Méndez, H., & Uribe, A. (1996). *Effect levels of nitrogen, phosphorus and potassium on Cocoa production in soils from the Department of Santander.*
19. Mattsson, B., Cederberg, C., & Blix, L. (2000). Agricultural land use in life cycle assessment (LCA), case studies of three vegetable oil crops. *Journal of Cleaner Production, 8*(4),283-292.
20. Medeiros, M. L., & Lannes, S. C. S. (2010). Propriedades físicas de substitutos do cacau. *Food Science and Technology, 30*(Supl. 1),243-253.

21. Mouron, P., Nemecek, T., Scholz, R. W., & Weber, O. (2006). Management influence on environmental impacts in an apple production system on Swiss fruit farms: Combining life cycle assessment with statistical risk assessment. *Agriculture, Ecosystems and Environment, 114*,311-322.
22. Ntiamoah, A., & Afrane, G. (2008). Environmental impacts of cocoa production and processing in Ghana: life cycle assessment approach. *Journal of Cleaner Production, 16*(16),1735-1740.
23. Ozdemir, G., & Maghsoodloo, S. (2004). Quadratic quality loss functions and signal-to-noise ratios for a trivariate response. *Journal of Manufacturing Systems, 23*(2),144-171.
24. Pardo, G., & Zufía, J. (2012). Life cycle assessment of food-preservation technologies. *Journal of Cleaner Production,* 28,198-207.
25. Phadke, M. S. (1989). *Quality Engineering Using Robust Design.* Englewood Cliffs: Prentice Hall.
26. Roer, A.-G., Korsaeth, A., Henriksen, T. M., Michelsen, O., & Strømman, A. H. (2012). The influence of system boundaries on life cycle assessment of grain production in central southeast Norway. *Agricultural Systems, 111,* 75-84.
27. Ruviaro, C., Gianezini, M., Brandao, F. S., Winck, C. A., & Dewes, H. (2012). Life cycle assessment in Brazilian agriculture facing worldwide trends. *Journal of Cleaner Production, 28*,9-24.
28. Suárez, Y. J., & Hernández, F. A. (2010). *Management of diseases of cacao (Theobroma cacao L) in Colombia with emphasis on Monilia (Moniliophthora roreri).*
29. United Nations Environment Programme - UNEP. Division of Technology, Industry, and Economics. Sustainable Consumption & Production Branch - SETAC. (2011). *Life Cycle Initiative Towards a Life Cycle Sustainability Assessment:* Making informed choices on products.
30. United Nations - UN. Commission on sustainable development - CSD. (2005).
31. Von Wirén-Lehr, S. (2001). Sustainability in agriculture - an evaluation of principal goal oriented concepts to close the gap between theory and practice. *Agriculture, Ecosystems & Environment, 84*(2), 115-129.

32. Zufia, J., & Arana, L. (2008). Life cycle assessment to eco-design food products: industrial cooked dish case study. *Journal of Cleaner Production, 16*(17), 1915-1921.

Chapter 3

Production and Quality Evaluation of Cocoa Products (Plain Cocoa Powder and Chocolate)

Ndife Joel[1], Bolaji Pius[1], Atoyebi Deborah[1] and Umezuruike Chris[2].

[1]Department of Food Technology, Kaduna Polytechnic, Nigeria.

[2]Department of Food Science and Technology, Michael Okpara University of Agric.,Umudike-Umuahia, Nigeria.

ABSTRACT

Raw cocoa beans were used to produce cocoa powder and chocolate samples. The processing steps include fermentation, drying, roasting, milling, pressing and conching. The fermentation time varied from 1 to 7days, represented as samples A, B, C, D, E, F and G respectively, in order to monitor the influence on the moisture content and cocoa products quality. The reconstition-characteristics, chemical composition and sensory quality of the cocoa powder and chocolate samples were determined. The results show that the average moisture loss in the beans after roasting was 75.13%. The reconstition properties of cocoa powders show that the dispersibility ranged from 18.10% to 60.50%, bulk density ranged from 0.69 to 0.83g/cm^3 and rehydration time ranged from 19 to 54 seconds. While the average proximate composition for cocoa powder and chocolate were: moisture (6.20% and 5.62%), fat (11.28% and 32.81%), protein (8.14% and 6.80%), fibre (1.80% and 2.59%), ash (5.81 and 2.15), carbohydrate (61.74% and 43.97%) and energy (498.39 Kcal) respectively. The organoleptic evaluation on the chocolate samples showed that there were significant differences ($p<0.05$) in the sensory attributes of colour, aroma, texture and taste. Sample GC derived from seven days fermented raw cocoa-beans was adjudged the best based on the overall acceptance.

INTRODUCTION

Cocoa beans are derived from the fruit of the plant Theobroma cacao L. In Nigeria, dry cocoa beans is majorly exported as a foreign exchange earner, while a small percentage of the cocoa beans serve as raw material for cocoa powder, cocoa butter and chocolate products (Adeyeye et al, 2010). Cocoa as a food ingredient is fast becoming very popular in the food and confection industry worldwide. It is available in a wide variety of forms, colors and flavors and is used in numerous applications (Borchers et al, 2000). A good quality cocoa powder should be relatively free flowing, stable and uniform in colour and flavour, of good microbiological quality, and easy to handle by the user (Vu et al, 2003). Moreover, a range of other characteristics such as pH, fineness, fat content, wettability, solubility and dispersibility, define the powder and have an important impact on the end product for which the cocoa is used (De- Muijnck, 2005).

The nutritional quality of cocoa products are determined largely by the chemical composition of the cocoa powder, which is dependent on the quantum of proteins, carbohydrates, fats, minerals and phytochemicals in the cocoa products and the corresponding digestibility coefficient (Belscak et al, 2009; Adeyeye et al, 2010; Lettieri-Barbato et al, 2012). Cocoa beans, as well as cocoa derived products, also present a rich source of phytonutrients, particularly catechins and procyanidins (Lecumberri et al, 2007). The total poly-phenol content of the bean is estimated to be 6-8% by weight of the dry bean (Wollgast and Anklam, 2000). Cocoa polyphenols have been reported in many studies as bioactive compounds, with antioxidant, antiradical and anticarcinogenic properties (Counet et al, 2006;Othman et al, 2007; Belscak et al, 2009; Lettieri-Barbato et al, 2012).

Cocoa powder is used in making beverages with other ingredients such as milk and sugar while cocoa butter is used for chocolate production. Chocolate products are desired and eaten, due to their attractive flavours and appearances (Othman et al, 2007; Pimentel et al, 2010). The primary chocolate categories are dark, milk and white (Afoakwa et al, 2007). The widely enjoyed chocolate-flavour, make it a favorite ingredient in bakery, ice cream, beverage, syrup manufacture and as a confection in itself (Lecumberri et al, 2007). Nowadays, consumers are more concerned with the nutritional status of foodstuffs and considering that cocoa powder and chocolate are extremely rich sources of many essential nutrients and phytochemicals that can contribute to a healthy diet (Lecumberri et al, 2007; Ieggli et al, 2011) highlight renewed interest in such products. The objectives of this research therefore, were to produce cocoa products (plain cocoa powder and chocolate) from raw cocoa beans and to assess the nutritional and sensory quality of the cocoa products as well as the consumer overall acceptance.

MATERIALS AND METHODS

The fresh cocoa beans were obtained from a local cocoa farm center in Akure, Ondo state, while, ingredients such as; sugar, nut-meg and milk were purchased from the central market, Kaduna in Nigeria. Production of plain cocoa powder and chocolate: Fresh cocoa pods were harvested ripe and opened to extract the wet beans. The wet beans were scooped with table knife from the pod into the basket. The basket containing the beans were initially lined and covered with banana leaves. The wet beans with their pulps were allowed to ferment for 7 days. The wet beans were labeled as samples A, B, C, D, E, F and G for each day of fermentation. The daily ambient temperatures and moisture contents of the fermenting beans were observed and recorded. After which the fermented pulps and other extraneous materials like sand-residues, chaffs, beans shell and hollow beans were manually removed and the beans were washed to remove the remnant mucilage. The fermented cocoa beans were dried in hot air oven at a temperature of 600C for 23 hours for each day of fermentation and their moisture contents were analyzed. The different fermented dried beans were roasted at same temperature of 1200°C for 65 min.

The roasted beans were cracked (kibbled) after which the shells were separated from the roasted cotyledons by winnowing. Most of the Nibs (90%) for each sample were ground using a grinding machine, to give cocoa mass and the remainder (10%) was left for subsequent chocolate production. 150g of cocoa mass were separated from each sample of cocoa mass. The remaining cocoa mass samples were pressed using a hydraulic-press and muslin cloth in order to extract the fat (cocoa-butter). The dried pressed cake for each sample were ground and sieved to obtain desired size particles of cocoa powder and labeled AP, BP ,CP, DP, EP, FP and GP respectively. The cocoa mass (cocoa butter and nibs) were mixed with other ingredients of sugar, milk and nutmeg. The cocoa-mass-mix for each sample was conched at 800C for 45min, using a stone-mill to give velvet smoothness. Each of the samples was tempered by stirring and cooling to 45°C and was poured into molds for subsequent hardening and shaping. They were wrapped with aluminum foil and labeled as samples AC, BC, CC, DC, EC, FC, and GC, after which they were stored in a freezer at -5⁰C, from where samples were taken for further analyses.

Physico-Chemical Analysis

The chemical composition of the cocoa powder and chocolate samples viz: pH, moisture, protein, fat, crude fiber and ash contents were determined by methods described by AOAC (1990). Carbohydrate was calculated by difference, and energy was calculated using Atwater conversion factors. The reconstition characteristics of the cocoa powder samples, such as bulk density, dispersibility and rehydration in both hot water and cold water were carried out as described by Compaore et al, (2011). Sensory analysis: Sensory evaluation of the chocolate samples were carried out by 25 panelists of judges, on a 9 point hedonic scale for different parameters such as colour, aroma, taste, texture and overall acceptability as described by Iwe (2010). Statistical analysis: The sensory evaluation data was statistically analyzed using the analysis of variance (ANOVA) and the Duncan Multiple range test with significance level at $p<0.05$ (Ihekoronye and Ngoddy, 1985).

RESULTS AND DISCUSSION

Moisture Content Assessment of Cocoa Beans

The moisture contents of raw beans (unfermented), dried beans (after fermentation), roasted beans and the final percentage moisture loss of cocoa beans are shown on table1. The results show that fermentation time has some effect on the moisture content of the processed cocoa beans. The average moisture contents in raw, dried and roasted beans were 28.80%, 18.14% and 7.11% respectively. Asiedu (1989) had reported that the moisture content of processed cocoa beans would fluctuate depending on the fermentation and drying conditions. Moisture is necessary for the biochemical changes that may influence the sensory and shelf quality of the cocoa products.

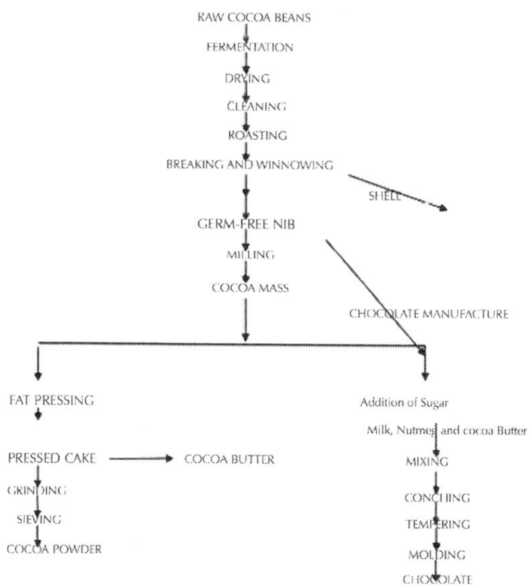

Figure1: Flowchart for the production of cocoa powder and chocolate.

Table 1: Moisture contents (%) of cocoa-beans during processing

Samples Fermentation Time	(days)	Raw	Dried	Roasted	%loss
A	1	25.28	16.36	6.88	72.79
B	2	27.30	17.24	7.10	73.40
C	3	30.45	18.42	7.84	74.25
D	4	35.37	20.10	8.15	76.96
E	5	31.25	22.25	7.33	76.54
F	6	26.40	17.15	6.52	75.30
G	7	25.53	15.49	5.92	76.68

Reconstitution-Ability of the Cocoa Powders

The results obtained from the reconstitution assessment of the cocoa powders are presented in Table 2. The bulk density is an important consideration in transporting, storing and packaging particulate materials (Onwuka, 2006). The bulk density of cocoa powders is affected by their moisture contents (Asiedu, 1989). The bulk densities seem to increase as the percentage moisture loss in most of the cocoa powder samples. Sample GP with high moisture loss (76.68%) recorded a bulk density of 10.69(g/ml). The average dispersibility was 42.01% for the cocoa powders, with sample GP having the highest dispersibility of 60.50%. Dispersibility describes the ease with which the cocoa powder may be distributed, as single particles, over the surface and throughout the bulk of the reconstituting water. Dispersibility is reduced by clump formation and is improved when sink-ability is high (Compaore et al, 2011). Substances such as starch or potassium bi-carbonate (alkali) may be added to cocoa powders to prevent caking, neutralize the natural acids and astringents, with the purpose of improving its dispersibility (Compaore et al, 2011). The wet-ability, that is, the wetting time provides useful indication of the degree to which the cocoa powders are likely to posses instant characteristics (Onwuka, 2006). The results suggest that the shorter the fermentation time the longer the wettability of the cocoa powders. Sample AP, took the longest time to get wet in cold water (54sec.) and hot water (32sec.). On the average the wettability in all the cocoa powders were better in hot water (34sec.) than cold water (19sec). The rates of biochemical reactions are faster at higher temperatures than at lower temperatures (Compaore et al, 2011).

Table 2: Reconstitution Assessment of Cocoa Powder

Parameters	A_p	B_p	C_p	D_p	E_p	F_p	G_p
Bulk density (g/ml)	0.72	0.73	0.81	0.79	0.83	0.78	0.69
Dispersibility (%)	18.10	25.23	37.35	46.15	50.30	56.4 2	60.50
Wet-ability in cold water (sec)	54	45	33	38	29	23	19
Wet-ability in hot water (sec) (85°C)	32	25	22	31	20	14	12

Chemical Analysis

The result obtained from the proximate analysis of the cocoa powders is shown in Table 2. The pH of the cocoa powders increased with the fermentation time. Sample AP had the lowest value of 3.91while sample AG had the highest value of 5.92. Adeyeye et al (2010) reported that the pH of well fermented and dried West African beans is around 5.5 whilst those of unfermented or poorly fermented beans are 5.0 or less. Fermentation normally involves the conversion of pulp sugar to alcohol by yeasts and alcohol to acetic acid by lactic acid bacteria as the fermentation progress. The acidity is reduced by the draining of citric acid during the sweating of the cocoa beans. The pH of the chocolate samples also increased as in the cocoa powders, with ranges from 5.65 to 6.15 for samples AC to GC. Changes in the acidity and the subsequent 0.69 drying kill the beans and help to develop the sensory characteristics of the cocoa powder and chocolate products (Asiedu, 1989; RodriguezCampos, 2012).

Fermentation helps to generate proper aroma and reduce the level of acetic acid, which causes off-flavour in chocolate (Adeyeye et al,

2010; Rodriguez-Campos, 2012). The drying reduced the moisture content of the cocoa powders to an average value of 6.20%. The longer the fermentation time the higher the final moisture contents in both the cocoa powders (samples AP to GP) and chocolates (samples AC to GC) with ranges of 5.12% to 7.20% and 5.15% to 6.23% respectively. The moisture content values of both the cocoa powder and chocolate products fall within the standard range to reduce the eventual growth of both bacteria and moulds and improve the shelf stability of the products (Guehi et al, 2010).The chocolate samples underwent further processing (conching) which was responsible for the reduced moisture contents when compared to that of cocoa powders (Borchers et al, 2000). Cocoa powder samples had higher range of protein values (6.80% to 9.55%) when compared to the protein values obtained for chocolate products (6.10% to 7.37%). This may be as a result of the conching process (800C for 45mins) which could have denatured some protein in the chocolate. The fat contents of cocoa powder samples ranged from 10.05% to 12.65%. Sample AP had the highest percentage fat content (12.65%) compared to sample GP (10.05%).While the fat contents for the chocolate samples ranged from 31.25% to 35.10%. The significant increase in the fat content of chocolate samples over that of cocoa powder samples was as a result of the contribution of ingredients added in the production of chocolate such as cocoa butter, milk and nutmeg. Fats, especially the unsaturated fat are prone to oxidation and shorten shelf-life of food products (Borchers et al, 2000; Afoakwa et al, 2007).

It was observed that the crude fibre contents of the both the cocoa powder and chocolate samples decreased with the fermentation time. Also the average crude fibre values for cocoa powder (1.80%) were lower than for chocolate samples (2.60%). This may be due to presence of higher soluble residues from the cocoa powder resulting from fermentation and not captured in the crude fibre analysis. Also, the direct use of cocoa nibs in the production of the chocolates must have contributed to the higher fibre content. Redgwell et al, (2003) showed the dietary fibre content of cocoa products to increase after roasting and conching, possibly due to the interaction between polysaccharides, proteins, polyphenolics and Maillard products at high temperatures.

The average ash content of cocoa powders (5.81%) was relatively higher than in chocolate samples (2.15%). Ash is an indication of mineral contents of foods and has been shown by leggli et al (2011) to be high in cocoa products. Afoakwa et al, (2007) reported that chocolates are good sources of minerals, specifically potassium, magnesium, copper and iron. There was significant increase in the carbohydrate value of cocoa powder (43.92%) compared to that of cocoa chocolate (61.74%), this could be as a result of ingredients added such as cocoa butter and milk, which were high in their fat and protein contents (Borchers et al,2000). The average energy value of the chocolate samples was 498.39Kcal. The high energy values of the chocolate samples can be explained by the high content of carbohydrates and lipids. This is desired especially by the growing and energetic young adults and because low-energy foods tend to limit the optimal utilization of other nutrients through the protective effect of carbohydrate on protein and polyphenols (Belscak et al, 2009; Lettieri-Barbato et al, 2012).

Table 3: Chemical Analysis of Cocoa Powder

Constituents	A_p	B_p	C_p	D_p	E_p	F_p	G_p
pH	3.91	4.45	4.98	5.10	5.36	5.61	5.92
Moisture %	5.12	5.58	5.75	6.26	6.70	6.91	7.10
Fat %	12.65	12.18	11.76	11.13	10.85	10.34	10.05
Protein %	6.80	6.96	7.38	8.50	8.82	8.94	9.55
Crude fibre %	2.64	2.28	2.15	1.78	1.40	1.21	1.06
Ash %	6.41	6.08	5.92	5.78	5.65	5.50	5.32
Carbohydrate %	62.47	62.48	62.06	61.45	61.22	61.49	61.00

Table 4: Formulation of ingredients for chocolate production

Ingredients (g)	A_c	B_c	C_c	D_c	E_c	F_e	G_c
Cocoa powder	350	350	350	350	350	350	350
Cocoa nibs	50	50	50	50	50	50	50
Cocoa butter	100	100	100	100	100	100	100

Sugar	250	250	250	250	250	250	250
Milk	250	250	250	250	250	250	250
Nutmeg	1	1	1	1	1	1	1

Table 5: Chemical analysis of chocolate

Constituent	A_c	B_c	C_c	D_c	E_c	F_e	G_C
pH	5.65	5.73	5.88	6.12	6.25	6.32	6.51
Moisture %	5.15	5.28	5.35	5.67	5.80	5.86	6.23
Fat %	31.25	32.30	32.79	31.78	32.63	33.84	35.10
Protein `)/0	6.10	6.39	6.85	6.88	7.15	7.19	7.37
Crude fibre %	3.16	2.85	2.65	2.48	2.41	2.32	2.24
Ash %	2.15	1.97	1.88	2.46	2.19	2.14	2.23
Carbohydrte %	46.54	45.48	44.60	44.61	43.57	42.33	40.32
Energy value	491.81	498.18	500.91	491.98	496.55	502.64	506.32

Sensory Evaluation

Table 7, present the result of the sensory evaluation carried out on the chocolate samples. The results for taste and aroma show similar score trend. Sample AC had the least scores for taste (6.42) and aroma (4.80), while sample GC had the highest score of (8.36) and (7.68) for taste and aroma respectively. There were significant differences between the samples. The panelists remarked on the astringent flavour in samples AC, BC and CC. This was probably as a result of shorter days of fermentation of the cocoa beans (RodriguezCampos, 2012). The pH and acidity of fermented cocoa beans were reported in different researches to influence both the taste and flavour of the products (Adeyeye et al, 2010). Asiedu, (1989) reported that the quality of the beans, which originally have a strong bitter taste, depends the fermentation, if it is overdone they may be ruined, if underdone they may have a raw flavors. However, the sensory scores for colour was converse to that of taste, aroma and texture. Sample AC had a higher score (7.92), than the other samples, while sample GC had the lowest score (5.54) for colour. It appeared that the panelist preferred the light-brown colours exhibited by samples GC, FC, EC to the dark-brown colours of AC, BC, CC and DC. Fermentation of cocoa beans used in the production, resulted in progressive darkening of the chocolates (Asiedu, 1989; Rodriguez-Campos et al, 2012).

The seven days fermented cocoa beans were also observed to be darker than the three days fermented beans this may be due to prolonged enzymatic browning during the seven days (Afoakwa et al, 2007; Rodriguez-Campos et al, 2012). All the chocolate samples had high scores for texture sensory attribute with an average value of 6.62. The score for sample GC (6.32) was better than the score for sample AC (7.10). The texture rating for samples DC, EC, FC, and GC were significantly higher than that of AC, BC and CC. The fermentation of cocoa beans was reported by Asiedu (1989) to result in softer cocoa products. The textural quality of finished chocolate products depends on the melting and crystallization behavior of the cocoa butter used (Olga and Biljana, 2004). Do et al, (2011) also showed that the chocolate produced from cocoa mass had a higher viscosity than chocolate produced from cocoa powder of the same total fat content. The average overall acceptability scores for all the chocolate samples were high (6.76). Sample GC derived from the beans with longest fermentation time of 7 days had the highest acceptability score (7.27), while sample AC, with 1 day fermentation time, had the lowest score (6.07).

There were significant differences between the chocolate samples for the sensory parameters at 5% significant level. Differences in the overall sensory perception of chocolate can be attributed to use of different fermented cocoa beans, functional variations in cocoa powder, other ingredient proportions and processing techniques (Afoakwa et al, 2007).

Table 6: Summary of sensory analysis for chocolate Sample Taste Colour Texture Aroma Overall acceptability

Sample	Taste	Colour	Texture	Aroma	Overall acceptability
A_c	5.12[5]	7.92[a]	6.32[5]	4.80[5]	6.07[5]
B_c	5.48[5]	8.10[a]	6.28[5]	5.28[5]	6.24[5]
C_c	6.16[5]	7.66[a]	6.34[5]	6.48[5]	6.66[5]
D_c	7.20[5]	6.80[5]	6.70[a]	6.72[5]	6.86[5]

E_c	7.48^5	6.44^5	6.75^a	7.72^a	7.10^8
F_c	8.12^a	6.14^5	6.82^a	7.65^a	7.15^a
G_c	8.36^a	5.54^5	7.10^a	7.68^a	7.27a

Means not followed by the same letters are significantly different at p (<0.05)

CONCLUSION AND RECOMMENDATIONS

In this study, the fermentation time of raw cocoa beans had influence on the physico-chemical and reconstitution properties of both the cocoa powders and their chocolate products. Seven days fermentation of raw cocoa beans was the best because it improved the nutritive quality and the sensory acceptability of the chocolate produced.

REFERENCES

1. Adeyeye, I.E., Akinyeye, O.R., Ogunlade, I., Olaofe, O.and Boluwade O. J. (2010). Effect of farm and industrial processing on the amino acid profile of cocoa beans. Food Chemistry, (118): 357-363.
2. Afoakwa, E. O., Paterson, A. and Fowler, M. (2007). Factors influencing rheological and textural qualities in chocolate – a review. Trends in Food Science & Technology, 18; 290-298.
3. AOAC (1990). Official Methods of analysis. 15th ed. Association of Official Analytical Chemists. Washington, DC.
4. Asiedu J.J. (1989) processing tropical Crops-a technological approach. Macmillan Education Ltd, London and Basingstoke, pp. 24-42.
5. Belscak, A., Komes, D., Horzic, D., Ganic, K. and Damir, K. D. (2009). Comparative study of commercially available cocoa products in terms of their bioactive composition. Food Research International, 42: 707-716.
6. Borchers, A. T., Keen, C. L., Hannum, S. M., and Gershwin, M. E. (2000). Cocoa and chocolate:composition, bioavailability, and

health implications. Journal of Medicinal Foods, 3:77-103.

7. Compaore, R.W., Nikiema, A.P., Bassole, N.H., Savadogo, A., Hounhouigan, D., Mouecoucou, J.and Traore, A.S. (2011). Nutritional properties of enriched local complementary flours. Advance Journal of Food Science and Technology, 3(1): 31- 39.

8. Counet, C., Callemien, D. and Collin, S. (2006). Chocolate and cocoa: New sources of transresveratrol and trans-piceid. Food Chemistry, 98: 649-657.

9. De Muijnck, L. (2005) Cocoa. In: Encapsulated and Powdered Foods (Onwulata, C., ed.), CRC Press, Boca Raton, USA.

10. Do, T.A., Vieira J, Hargreaves, M.J., Mitchell, R.J. and Wolf, B. (2011). Structural characteristics of cocoa particles and their effect on the viscosity of reduced fat chocolate. LWT - Food Science and Technology, 44:1207-1211.

11. Guehi, T.G., Zahouli, I.B., Ban-Koffi, L., Fae, M.A. and Nemlin, J.G. (2010). Performance of different drying methods and their effects on the chemical quality attributes of raw cocoa material.

12. International Journal of Food Science and Technology, 45: 1564-1571.Am. J. Food. Nutr, 2013, 3(1): 31-3838

13. Ihekoronye, A.I and Ngoddy, P.O. (1985). Integrated food science and technology for the Tropics.Macmillan publishers, London. pp. 310-316.

14. Iwe, M.O. (2010). Handbook of Sensory Methods and Analysis. Rojoint Communication Services Ltd.,Enugu. pp. 75-78.

15. Lecumberri, E., Mateos, R., Izquierdo-Pulido, M., Ruperez, P., Goya, L., La Bravo, L. (2007). Dietary fibre composition, antioxidant capacity and physico-chemical properties of a fibre-rich product from cocoa (Theobroma cacao L.). Food Chemistry, 104: 948-954.

16. Lettieri-Barbato, D., Villano, D., Beheydt, B., Guadagni, F., Trogh, I. and Serafini, M. (2012). Effect of ingestion of dark chocolates with similar lipid composition and different cocoa content on antioxidant and lipid status in healthy humans. Food Chemistry, 132:1305-1310.

17. Ieggli, C., Bohrer, D., Nascimento, P. and Carvalho, L. (2011). Determination of sodium, potassium, calcium, magnesium, zinc and iron in emulsified chocolate samples by flame atomic

absorption spectrometry. Food Chemistry, 124:1189-1193.

18. Othman A, Ismail A, Abdul-Ghani N and Adenan I (2007). Antioxidant capacity and phenolic content of cocoa beans. Food Chemistry, 100:1523-1530.
19. Onwuka, G.I (2005). Food Analysis and Instrumentation: Theory and practice. Naphthali Prints Lagos, Nigeria.
20. Olga Jovanovic and Biljana Pajin (2004). Influence of lactic acid ester on chocolate quality. Trends in Food Science & Technology, 15:128-136.
21. Pimentel, F., Nitzke, J., Klipel, C. and Vogt de Jong, E. (2010). Chocolate and red wine – A comparison between flavonoids content. Food Chemistry,120:109-112.
22. Redgwell, J.R., Trovato,V. Curti, D. (2003). Cocoa bean carbohydrates: roasting-induced changes and polymer interactions. Food Chemistry, 80: 511-516.
23. Rodriguez-Campos, J., Escalona-Buendía, H., Contreras-Ramos, S., Orozco-Avila, I., JaramilloFlores, E. and Lugo-Cervantes ,E. (2012). Effect of fermentation time and drying temperature on volatile compounds in cocoa. Food Chemistry, 132:277-288.
24. Wollgast, J. and Anklam, E. (2000). Polyphenols in chocolate: is there a contribution to human health.? Food Research International, 33:449-459.
25. Vu, T.O., Galet, L., Fages, J. and Oulahna, D. (2003). Improving the dispersion kinetics of a cocoa powder by size enlargement. Powder Technology, 130(1-3): 400-406.

Chapter 4

Adoption of New Technologies of Traditional Transformation of Products Made from Cocoa (Theobroma Cacao L.) in Cameroon

J. Amang A Mbang[1], P. Mounjouenpou[1], E. Nossi[2], D. Achukwi[1], S. Bassanaga[3], and N. Woin[1]

[1]Institut de Recherche Agricole pour le Développement (IRAD), BP 2067, Yaoundé, Cameroun

[2]Université de Yaoundé 1, Département de Sociologie, BP 453 Yaoundé-Cameroun

[3]Groupement d'Initiative CONACFAC, B.P. 34434 Yaoundé –Cameroun

ABSTRACT

The project "Training and monitoring of women's groups on the manufacture of cocoa products" of World Cocoa Foundation (WCF) with the Institute of Agricultural Research for Development (IRAD) and NGO's CONAFAC, have permitted to diffuse some new technologies of cocoa processing into derived products among farmers in order to reduce poverty, malnutrition and food insecurity in rural areas. The objective of this study is to measure the level of utilization of new technologies and to determine factors that can affect the adoption of these new technologies by farmers. The methodology used in data collection was the direct survey. The levels of adoption were confirmed by a descriptive analyses method, probate and generalized linear model. A case study was done in Mbalmayo and Mbangassina. The results showed that, generally, cocoa butter was the most adopted innovative product, followed by cocoa power and soy-chocolate drink. The adoption rate may differ with studied areas. The major difficulties encountered during the technological adoption were as follows; crushing cocoa bean (30%), lack of cocoa bean (21%), lack of adequate sunshine (12%), lack of electricity (11%). The econometric Logit Model showed that two variables; occupation and level of education influenced the adoption model positively and significantly (coefficient of regression are 0.54806 and 0.16477 respectively). Age group (-0.04778), Marital status (-0.00094) and sex (-0.14214) were negative with no significant coefficient of regression. Our results indicate that cocoa bean transformation into its varied derivatives, offer opportunities to farmers to generate more income. But most female cocoa farmers do not own personal cocoa farm land, and do not have adequate processing materials.

Both factors have negative effect on the adoption of new technologies.

INTRODUCTION

In 2005, the population of Cameroon was estimated at 17,463,836 inhabitants with an average growth rate of 2.8% / year. According to the Central Bureau of Census and Population Studies [1], the population is expected to reach 26.5 million in 2020, with an estimated average density of 37.5 habt / km ^{2}. This density changes geographically: there are regions with low average and those with high population density. Women constitute about 50.5 % of the population [2]. Over the years, Cameroonian agriculture faced the challenges of feeding a rapidly growing population. She will have to sufficiently increase food production and income to ensure food security in Cameroon and Africa. Researches in agriculture sector are quite challenging and enormous: it entails to favor an increase in crop yields, define, improve and transfer new technologies to stakeholders. The challenge is also to improve food quality for better marketing and competitiveness. All of these changes must be environmentally friendly.

World cocoa production goes from 18 000 tons in 1850 to 155,000 tons in 1900 and reached 672 000 tons in 1940 [3, 4]. In 1997, cocoa production was close to 207 million tons.

Today it is 3.5 billion tons and Africa alone provides 72 % of the production. Cameroon cocoa production is estimated at 220 000 tons in recent years [4] and ranks 4th in Africa.

Fermented and dried cocoa beans, is the raw material used for the grinding industry. After roasting and removing the shell, it is used to manufacture semi-finished products (cocoa paste and its direct derivatives: cocoa powder and cocoa butter), or finished products intended for direct consumption (chocolate powder, tablets or chocolate confectionery). Thus women found in these two regions of cocoa production (Mbalmayo and Mbangassina) in the Central region were trained on the traditional transformation of cocoa beans into its derivatives: cocoa butter, cocoa powder and soy chocolate drink, with the aim of enable them practice theses technologies.

Objective

The objective of this study is to analyze the perception of women cocoa farmers' as concerns the processing of cocoa into its derivatives as well the difficulties encountered during the adoption technology.

Most specifically, this work would aim at;

- Estimating the level of adoption and diffusion of technologies;
- Determining the socio-economical and institutional factors influencing technological adoption.

METHODOLOGY

Study Areas

Two important pools of Cameroonian cocoa production have been selected for this study: Mbangassina and Mbalmayo. Mbangassina is a highly agricultural and far off division, whereas Mbalmayo is a more developed sub division along the main road between Yaoundé and Kye-ossi (border between Cameroon, Gabon and Central Africa).

Surveys were conducted in 6 villages of Mbangassina and 5 villages of Mbalmayo. These two areas belong to two different agro-ecological zones: forest zone (Mbalmayo) and transition zone (Mbangassina) (Figure 1).

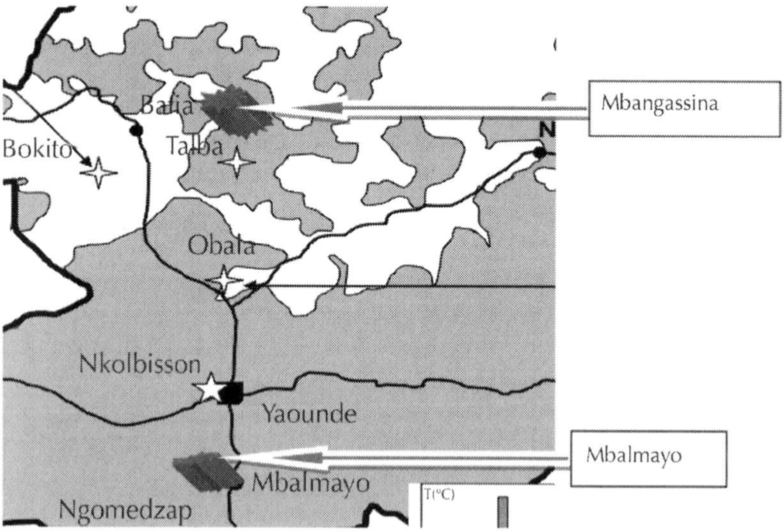

Figure 1: Localization of studied zones.

Technology of Cocoa Processing

Women cocoa farmers of Mbangassina and Mbalmayo regions were trained on the manufacture of cocoa butter, cocoa powder and soy chocolate drink (figure 2). This training took place at a selected study zones. The technologies to be transfer were simple and practicable, using local materials that were available in rural areas (pot, tissue, spoon, fire wood, fire side, clean water, mortar and pistil). These technologies were defined by Mounjouenpou and al [5].

Cocoa Butter: Cocoa beans were placed in a locally made aluminum heating pot for roasting and were regularly being stirred for about 60 minutes under the temperature of about 50°C. Next, the roasted cocoa beans was introduced into a mortar and lightly pounded in order to facilitate removal of the husk by winnowing. Thereafter, the beans were finely ground into paste with the aid of a mill. Water was added to the paste and then cooked for several hours. During boiling, cocoa butter emerged to the surface of the mixture, and was collected with the help of a spoon. This liquid cocoa butter was washed using clean water and filtered using a clean tissue.

Cocoa Powder: After extracting the cocoa butter, cocoa paste was obtained by evaporating the water. Cocoa paste was dried to obtain cocoa cake. The dried cake was finely ground unto the desired consistency. Cocoa powder obtained can be consumed as a hot drink with sugar or honey during breakfast or as cold chocolate drink.

Soy-chocolate Drink: To prepare 3 liters of this drink, soak 500 g of soya-beans in a saline solution for 24hrs. Whiten the soaked grains. In order to eliminate the smell of the beans, boil for 10 to 15 minutes at 80°C. Grind the soy-beans with little water. Measure 300g of soybeans paste that will be poured into a large saucepan. Add 3 liters of water and 30% of cocoa paste (approximately 100g). Filter well with a very fine mesh sieve. After grinding, the collected paste is boiled for 10 to 15 minutes. The drink is flavored by adding lemongrass and later sweetens to give the desired taste.

Two years after the transfer of technologies to stakeholders, researchers organized and supervised adoption surveys in different areas where women were being trained. Techniques and adoption tools for survey were being defined according to the village, area and the condition of lifestyle of households. Generic questionnaires were being made to collect data.

Enquiries focused on women who were trained by researchers and those who were trained by "trainer of trainers". The adoption survey concern 127 women of two selected regions:

Cocoa butter

Cocoa powder

soy-chocolate drink

Figure 2: Cocoa butter, cocoa powder, and soy chocolate drink.

Data Collection and Analysis

Primary data was collected through an area of study by a survey to obtain the characteristics of the farmers and the level of their adopting the new technology. The structured survey served as a control for checking or comparing information obtained through participatory methods. Information was gathered through a questionnaire administered to the farmers by the facilitators.

Statistical analysis were performed on an SPSS (release 16.0)[6] computer package, Logistic Model to explain the ad optional factors, with coefficients of regression and procedure in SAS 9.2 (SAS Institute, 2002)[7] with the DUNCAN test to separate the mean.

Analytical Model

The decision to adopt a technology is dichotomous; a woman can decide whether to use the technology or not. The adopter was defined as a woman who uses one of these technologies regardless of the frequency. The decision to adopt is considered as a dependent qualitative variable with a regression value of 0 or 1. This value depends on the characteristics of the adopter. The approach used in analyzing the factors determining the adoption can be estimated using a model which can predict the decision of farmers in the adoption of the technology proposed or not. The decision will also depend on the socio-economic characteristics of the decision maker. In other words, this model aims, at a given economic agent, to determine the probability to adopt or not to adopt a given technology. The literature on adoption studies allows for the distinguishing of at least three types of commonly used models to analyze the decision to adopt an agricultural technology: Linear Probability Models are Logit and Probit but the most commonly used model that best explains the diffusion process is the Logistic Function [8](Rogers, 1983). The Probit model has drawbacks because the probability can often exceed 1 but does not hold same for the other two, who through any transformation maintains the estimated probability between 0 and 1. We used the Logit Model, often used in case studies of technology adoption for reasons of convenience [9].

The model can be presented by the following equation:

$$E(Y_i) = P(Y_i) = \frac{e^{\alpha+\beta X_i}}{1+e^{\alpha+\beta X_i}}$$

Therefore, the probability of non - adoption of technologies becomes

$$P(not-adoption) = 1 - P(Yi) = \frac{1}{1+e^{\alpha+\beta X_i}}$$

Where:

P (Yi) = probability of an individual i to adopt the technology; P (Yi) = 1 if the technology is adopted and 0 if the technology is not adopted.

e = the exponential function

Yi = the dependent variable; adoption technology

Xi = characteristic of individual i; represents the vector of explanatory variables (age, village, etc.).

β = parametric vectors to estimate where the sign permits the interpretation of results

α = constant

Model Specification

Based on the results of studies on the adoption of technologies, some socio-economic factors are considered instrumental in the adoption of technologies. This includes age, gender, education, occupation and marital status. These variables that characterize the socioeconomic status of women can be instrumental in the adoption of technological innovations.

In fact, women are considered adopting a technology when making use of this technology at whatever intensity. This adoption is influenced positively or negatively by socio – economic and technical factors like explain Kebede and al [9], Ibro and Bokar [10]. These hypotheses are greatly influenced by the following variables:

- *Age*: the adoption of technology can be highly influenced by the age of the women. The adoption of new technologies requires a certain level of risk associated with the decision to apply technology. Young producers are willing to take more risk compared to older producers.
- *Education*: this variable can be of great importance in the adoption of new technologies. Those with a higher level of education will easily embrace new skill, innovations and assess new technologies compared to those at a lower level. This variable could positively or negatively influence the adoption of technologies.
- *Gender*: women from time immemorial have been proven to easily access information on processing compared to men. Because of these factors alone they stand a better chance of facilitating the adoption of new technology compared to men. The gender variable negatively affects the probability of adopting a new processing technology.

- *Occupation*: occupation may influence adoption. Women who are involved in agriculture and who own cocoa farms have a higher probability of adopting a new technology in the processing of cocoa into its derivatives. The Adoption rate is the proportion of peasants that make use of this technology. Descriptive statistics (Average, Frequency, and Histogram) were used for data processing.

RESULTS AND DISCUSSIONS

Results are a summary of the investigations undertaken in the study sites. Analysis identifies variables that can affect the adoption of technologies of processing cocoa beans into cocoa butter, cocoa powder and soy-chocolate drink.

Analysis of Socio-economic Profile

This study analyzes the socio-economic characteristics of respondents under the following: age, occupation, education and marital status. The data is represented on table 1. Table 1 indicates that the occupation of farmers occupied a dominant proportion (71%). those with a primary and secondary school levels of education, occupied at least 98% of the attendants, the level of education are at least in secondary and primary school (98%), the most dominant age group were greater than 40 years of age with most of them being married (as the dominant marital status). These data are in accordance with those published in several scientific reports [11, 12, 13, 14].

Table 1: Household Socio-demographic characteristic of respondent's involved in the survey on transformation technologies of adoption on cocoa beans

Characteristic	Sub-division		Overall mean
	Mbalmayo	**Mbangassina**	
Household size			
females	83	41	62

male	3	0	1,5
Categories of occupation in %			
Agriculture	77	33	
merchant	4 3	57	
Education level frequency			
non formal education	1	0	0,5
Primary school	35	14	24,5
secondary school	46	27	36,5
university	1	0	0,5
Age category frequency responding			
18-21 years	8	1	4,5
21-25 years	2	2	2
25-28 years	4	0	2
28- 32 years	2	1	1,5
32-40 years	16	6	11
< 40 years	52	29	35,5

Adoption of Technologies

Surveys have permitted the determination of the level of adoption of new technologies. The highest adoption rate of technologies was recorded in Mbalmayo where the women have tried to apply these three processing technologies at least once. In Mbalmayo, Approximately 78% of women who have been trained adopted: 71% of cocoa butter, 81% of cocoa powder and 95% of soy- chocolate drink. The lowest rates were recorded in Mbangassina, with rates of 29%, 19% and 5% respectively for cocoa butter, cocoa powder and soy-chocolate drink (Figure 3). Where ever the region this study was being carried, farmers did express some difficulties in applying the technologies due to the lack of adequate processing equipment and difficulties in crushing beans.

Table 2: Awareness of new methodologies of transformation of cocoa bean

Variables	Category	percent
Acceptante of new technologies		100
Level of practice	No	6
	Yes	94
How to crush cocoa bean	On hand machines	81
	In electrical machines	18
	On hand stones	1
Level of used new methods	cocoa butter	54
	Cocoa powder	34
	soy - chocolate drink	1
Level of selling new products	cocoa butter	58
	Cocoa powder	30
	soy - chocolate drink	12

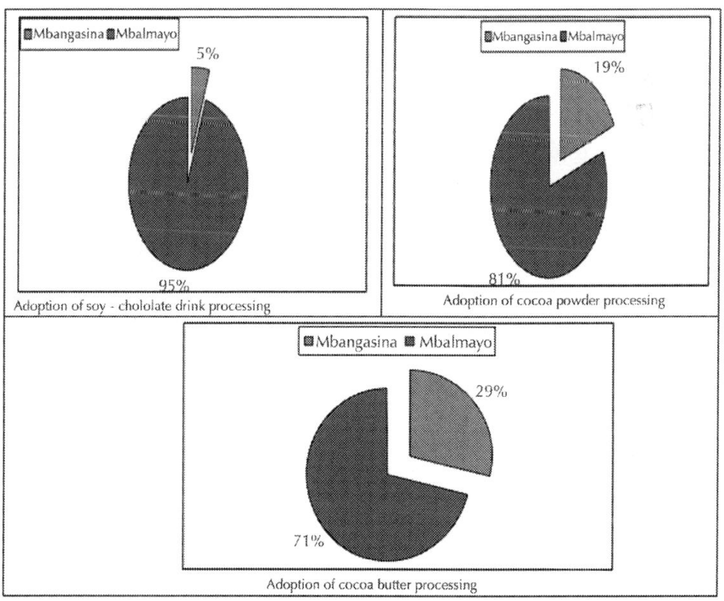

Figure 3: Level of adoption.

Factors Affecting Technologies Adoption

Table 3 provides econometric analysis of the adoption of technologies. Two variables namely socio-professional characteristics and the level of general education have a significantly positive influence on the adoption of processing technologies. The improvement in the adoption rate of new technologies therefore requires efficiency of the farming profession and an average level of education. Producers who went to school are aware of the importance of cocoa processing. On the contrary, variables based on: age, marital status and sex have negative regression coefficient and appears not to be significant on the use and adoption of cocoa processing technologies.

Table3: Results of the econometric analysis of the adoption of cocoa derived product technologies (regression coefficients)

Variables	P- Value
Occupation	0.54806 ***
Level of education	0.16477**
Age group (years)	-0.04778
Marital status	-0.00094
sex	-0.14214

, * : signifiant at 5%

Difficulties faced by farmers during the adoption of new technologies (table 4). The majors difficulties were: at the level of crushing cocoa bean (30%), lack of cocoa bean (21%), lack of electricity (11%) and lack of sunshine (12%).

Table 4: Farmers difficulties faced by farmers during the adoption of new

Constraints	Frequency	Percent
no water	1	2
Crushing cocoa bean	14	30
far of farm	1	2
remove skin of beans	1	2
roasting beans	2	4

4lack of electricity	5	11
lack of cocoa bean	10	21
smoked	1	2
lack of drying room	3	6
lack of sunshine	6	12
expensive of machine	1	2
presence of mousse	1	2
problem of conservation	1	2
problem of transport	1	2

In the two district (Mbalmayo and Mbangasina) the socio-characteristics and the adoption factors are not significant at $\alpha = 5\%$ level when we used the test of the comparison of DUNCAN in the procedure of GLM (Table 5). These confirm that the conditions of adoption of new technologies are the same in the entire study zone.

Table 5: Comparison of variables of two districts based on the generalize linear model

Variables	Districts		Over all means value	P-
	Mbalmayo	Mbangassina		
Sex	1.96512	2.00000	1.976378	0.2344
occupational category	1.2683	1.0513	1.198347	0.0754
Education level	1.4940	1.3415	1.443548	0.1873
Marital status	1.7647	1.7436	1.758065	0.8345
Age	5.0476	5.4615	5.178862	0.1479
Overall Innovative products	1.08235	1.05128	1.072581	0.5041
Cocoa butter	69.620*	60.811*	66.81034	0.0499

Cocoa powder	41.667	45.536	43.15068	0.1324
Soy-chocolate drink	30.048	29.167	29.85185	0.9204
Farmers difficulties when practicing the technology	3.7273	3.1250	3.609756	0.2878
Quantity of transformed cocoa butter per week	1.6042	1.7672	1.657303	0.7386
Quantity of transformed cocoa powder per week	3.60*	64.06*	21.15000	0.0197

* : significatif à 5%

Many scientists investigated on factors that could influence the acceptance of new technologies by households. The critical review of articles related to adoption of technologies has been done by Peek and al [15]. The majority of data available are based on qualitative data relative to adoption level. They have demonstrated that acceptance of new technology are influenced by 27 factors, divided into six themes: Concerns regarding technology (e.g., high cost, privacy implications and usability factors) [16, 17, 18];

- Expected benefits of technology (e.g., increased safety and perceived usefulness) [19,20, 21];
- Need for technology (e.g., perceived need and subjective health status) [19, 21];
- Alternatives to technology (e.g., help by family or spouse) [20, 22, 23];
- Social influence (e.g., influence of family, friends and professional caregivers) [19,23];
- And characteristics of older adults (e.g., desire to age in place) [24, 22].

CONCLUSIONS AND RECOMMENDATIONS

This study aimed at studying levels and constraints of adoption of new technologies of traditional transforming cocoa beans traditionally into its derivatives. A targeted population of 127 women cocoa farmers underwent training and monitoring on the manufacture of cocoa products: soy-chocolate drink, cocoa butter and cocoa powder. After field trip and inquiry, the adoption of transferred technologies was measured. The result portrayed that new technologies were mostly used by women of Mbalmayo than those of Mbangassina subdivision. Cocoa butter was the most adopted product, followed by cocoa powder. The major adoption constraints in the two areas are the lack of cocoa bean and some useful material. The lower percentage of adoption was recorded with soy-chocolate drink. The problem of conserving the drink due to the lack of electricity in the villages can also explain the low adoption rate of soy chocolate drink. To improve the adoption rate of these technologies by farmers, some constraints have to be solved; availability of cocoa, availability of the adequate material of transformation, reinforcement of local systems of communication.

ACKNOWLEDGEMENTS

We want to thank "World Cocoa Foundation (WCF)" for funding this study, all the cocoa farmers of Mbangassina and Mbalmayo regions, the team of IRAD researchers, CONACFAC NGO's and AWARD.

RÉFÉRENCES

1. BCREP, 2010 BCREP, 2010. Bureau Central de Recensement et d'Etude de la Population, rapport annuel de statistique de la population au Cameroun.
2. DSCE, 2009. DSCE, 2009. Document de Stratégie de la Croissance et de l'Emploi, Cameroun.http://www.afdb.org/fileadmin/uploads/afdb/Documents/Project-and Operations/Cameroon %20DSCE2009.pdf.

3. AmangA MbangJ. and al (2012) AmangA MbangJ. Mounjouenpou P., Babin R., Cilas C., Tchouamo I., Dibog L. & Nyassé S. 2011. Tentative prediction of cocoa yield (*Theobroma cacao* L.) in farmer's field in Cameroon by CART segmentation method. Journal of Natural product and plant resource (Vol.I, Issue 2).
4. ICCO. 2006. Etude de marché sur le cacao biologique. Cent trentième réunions du Comité Exécutif tenu à Londres du 1é au 15 septembre 2006.
5. Mounjouenpou P., Amang A Mbang J., Bernard Guyor, Joseph-Pierre Guiraud (2012). Traditional procedures of cocoa processing and occurrence of ochratoxin-A in the derived products, Journal of Chemical and Pharmaceutical Research: 4(2), 1332-1339.
6. SPSS version 16.0, 2007, SPSS Institute Inc.
7. SAS version 9.1, 2005. User's guide. SAS Institute Inc., Cary, NC USA.
8. Rogers, E.M., 1983. "Diffusion of Innovations" Third Edition Macmillan Publishers, LondonKebedeY.,K. Gunjal, and G. Coffin, 1990. Adoption of new technologies in Ethiopian Agriculture: The case of Tegulet-Bulga District, Shoa Province. Agricultural Economics, 4(1):Elsevier Science Publishers, Amsterdam.
9. Alpha K., M. Toure, O. Coulibaly et R. Adeoti 2001. Transfert de nouvelles technologies dans les systèmes de production des paysans au Mali. Adoption des variétés améliorées du Niébé. Rapport d'activités. PRONAF IITA, Cotonou, Bénin.
10. Ibro G., M. Bokar 2001. Transfert de nouvelles technologies dans les systèmes de production des paysans au Niger. Rapport d'activités PRONAF IITA, Cotonou, Bénin
11. DSRP, 2000. Document de Stratégie de Développement du Secteur Rural, Cameroun. http://camerounforet.com/fr/system/files/02_09_52.pdf.
12. DSDSR, 2005. Document de stratégie du développement du secteur rural, www.minepat.gov.cm/index.php?option=com_docman&task.
13. World Bank, 2010. Annual Report 2010. Washington D.C.
14. Food and Agricultural Organization (FAO), 2000. FAOSTAT 2000.
15. Sebastiaan T.M. Peek, Eveline J.M. Wouters, Joost van Hoof, Katrien G. Luijkx, Hennie R. Boeije, Hubertus J.M. Vrijhoef, 2014.

Factors influencing acceptance of technology for aging in place: A systematic review: International journal of medical informatics; 83, 235–248.

16. A. Mihailidis, A. Cockburn, C. Longley, J. Boger, Theacceptability of home monitoring technology amongcommunity-dwelling older adults and baby boomers, Assist.Technol. 20 (2008) 1–12.
17. K. Wild, L. Boise, J. Lundell, A. Foucek, Unobtrusive in-homemonitoring of cognitive and physical health: reactions andperceptions of older adults, J. Appl. Gerontol. 27 (2008)181–200.
18. J. Cohen-Mansfield, M.A. Creedon, T.B. Malone, M.J.Kirkpatrick III, L.A. Dutra, R.P. Herman, Electronic memoryaids for community-dwelling elderly persons: attitudes,preferences, and potential utilization, J. Appl. Gerontol. 24(2005) 3–20.
19. L.Lorenzen-Huber,M.Boutain,L.J.Camp,K.Shankar,K.H.Connelly, Privacy, technology, and aging: a proposedframework, Ageing Int. 36 (2011) 232–252.
20. C.K. Lai, J.C. Chung, N.K. Leung, J.C. Wong, D.P. Mak, A surveyof older Hong Kong people's perceptions oftelecommunication technologies and telecare devices, J.Telemed. Telecare 16 (2010) 441–446.
21. C.D. Steggell, K. Hooker, S. Bowman, S. Choun, S.J. Kim, Therole of technology for healthy aging among Korean andHispanic women in the United States: a pilot study,Gerontechnology 9 (2010) 443–449.
22. R. Steele, A. Lo, C. Secombe, Y.K. Wong, Elderly persons'perception and acceptance of using wireless sensornetworks to assist healthcare, Int. J. Med. Inform. 78 (2009)788–801.
23. G. Demiris, B.K. Hensel, M. Skubic, M. Rantz, Seniorresidents' perceived need of and preferences for smarthome sensor technologies, Int. J. Technol. Assess. HealthCare 24 (2008) 120–124.
24. K.L. Courtney, G. Demiris, M. Rantz, M. Skubic, Needingsmart home technologies: the perspectives of older adults incontinuing care retirement communities, Inform. Prim. Care16 (2008) 195–201.

Chapter 5

Incidence of Pod Integrity on the Fungal Microflora and Ochratoxin-A Production in Cocoa

Pauline Mounjouenpou[1], Dominique Gueule[2], Suzanne Abeline Maboune Tetmoun[3], Bernard Guyot[4], Angélique Fontana-Tachon[5], and Joseph Pierre Guiraud[6]

[1]Institute of Agricultural Research for Development (IRAD), Food technology laboratory PO box 2067, Yaounde, Cameroon

[2]CIRAD, UPR Qualité Produits Pérennes, TA80/16, 34398 Montpellier Cedex 5, France

[3]Institute of Agricultural Research for Development (IRAD), Food technology laboratory, PO box 2067, Yaounde, Cameroon

[4]CIRAD, UPR Qualité Produits Pérennes, TA80/16, 34398 Montpellier Cedex 5, France

[5]Université Montpellier II, UMR-IR2B (Agro-M/INRA/UM2), Place Eugène Bataillon, 34095 Cedex 5, France

[6]Université Montpellier II, UMR-IR2B (Agro-M/INRA/UM2), Place Eugène Bataillon, 34095 Cedex 5, France

ABSTRACT

Ochratoxin-A (OTA) is a mycotoxin that has nephrotoxic, tetragenic, immunotoxic and carcinogenic effects in the human organism. It contaminates several foodstuffs, notably cocoa. The purpose of our study was to compare the incidence of cocoa pod integrity on the fungal microflora and ochratoxin- A production in Cameroon. Irrespective of pod condition, fermented cocoa beans were contaminated by OTA. The maximum mould content was obtained in beans from damaged pod. However, the fungal microflora was more diversified for beans from pod damaged during harvesting than for beans from intact pods, throughout the post-harvest process. The toxigenic strains isolated belonged to the genera *Aspergillus*. *Aspergillus carbonarius* isolated from damaged pods, displayed the greatest OTA production which was 110.7 ng.g^{-1} on cocoa medium and 2772,0 ng.g^{-1} on official medium. There was a good correlation between OTA presence in the beans and isolated toxigenic strains.

INTRODUCTION

Ochratoxin A (OTA) is a toxic secondary metabolite produced by several species of *Aspergillus* and *Penicillium* genera. OTA is mainly produced by *Aspergillus carbonarius*, A. niger and A. ochraceus in tropical zones, and by *Penicillium verrucosum* and *P. nordicum* in temperate zones (Pitt et al., 2000; Abrunhosa et al., 2001; O'Callaghan et al., 2003).

OTA attracts particular attention due to the damage it causes in the human and animal organism (Abarca et al., 1998). It has nephrotoxic (Mantle and McHugh, 1993), immunotoxic, teratogenic and carcinogenic (Kuiper-Goodman and Scott, 1989; Kuiper-Goodman, 1996; Höhler, 1998) effects in human and animal organisms.

OTA contaminates several foodstuffs and drinks, notably cocoa (Pittet et al., 1996; Blanc et al., 1998; Hurst and Martin, 1998,

Jorgensen, 1998; Skaug, 1999; Gareis and Scheuer, 2000; Thirumala-Devi et al., 2001; WHO, 2001).

Fermentation is the main stage in cocoa post-harvest processing. It is generally carried out in a traditional manner under the action of natural microorganisms. Many studies have been carried out on the influence of post-harvest processing on OTA contamination in coffee (Suàrez-Quiroz et al., 2005, Durand, 2009; Duris et al., 2010) and cocoa (Mounjouenpou et al., 2011).

The purpose of our study was to compare the incidence of cocoa pod integrity on the fungal microflora and Ochratoxin- A production in Cameroon.

MATERIAL AND METHODS

Cocoa Post-harvest Treatments

Heap fermentation of cocoa pods (*Theobroma cacao* L.) from the Kumba region of Cameroon was done either immediately after harvesting in the field with undamaged or damaged pods, or 10 days later with damaged pods. Fermentation was carried out in each case during 5 days, and using 50 kg of beans. In doing heap fermentation, the beans were tipped into banana leaves placed on the ground. The heap was then covered with other banana leaves. Natural drying (in the sun) was carried out for between 5 and 10 days. Cocoa samples were taken at different stages of processing. This involved unfermented beans and fermented sun-dried beans.

Microbiological Analyses

The inoculum was obtained by soaking 15 cocoa beans in 90 ml of a peptone water solution (0.1% w/v) for 10 min (Hocking, 1991). The surface of PDA medium was inoculated and the dishes were incubated at 25°C for 5 to 7 days. Isolated moulds were set apart according to the identification key for common food-borne fungi (Samson et al., 1995). The identification of *Aspergillus* and *Penicillium* moulds was confirmed using molecular techniques by the Fungi and Yeasts Culture

Collection at the Catholic University of Louvain in Belgium (BCCMTM/MUCL Culture Collection).

STUDY OF OTA PRODUCTION BY ISOLATED STRAINS

PDA Culture Medium

The study of OTA production in PDA culture medium was described by Sùarez-Quiroz et al. (2004). For each strain isolated, a suspension of 3 x 10^6 of conidia.mL^{-1} was made up by scraping a PDA culture dish with a saline solution containing 0.01% Tween 80. Five microliters of the suspension was deposited in the centre of a dish of PDA medium which was incubated at 25°C. After 20 days of incubation, direct extraction was carried out from 3 agar discs taken from the centre of the colony. This extraction was done in 2.5 mL of solvent (methanol/formic acid 25:1, v/v) for 15 min in an ultrasound bath.

Cocoa Culture Medium

The method used to study OTA in cocoa medium are described by Mounjouenpou et al. (2008). In short, 50 g of cocoa beans (verified OTA-free) were inoculated with 8 mL of a suspension of 50 x 106 conidia.mL^{-1} and incubated at 25°C for 20 days. Extraction was carried out in an acetonitrile/water solution (60:40 v/v) for 40 min.

Rice culture Medium (FDA method)

The FDA method is the official method to study the capacity to produce OTA by moulds. This method is described by Tournas et al., (2001).

In all cases, OTA was quantified on extracts by HPLC with fluorimetric detection (Shimadzu LC-10 ADVP, Japan) (Nakajima et al., 1997). The operating conditions were as follows:

100 μl injection loop, C18 reverse phase HPLC column, ODS 5 μm with an identical pre-column thermostatically controlled at 35°C,

an isocratic flow of 1ml/min, an excitation wavelength of 333 nm and an emission wavelength of 460 nm. Contents were calculated from a calibration curve established from the standard of 1µg.mL^{-1}; ref PD 226 R. Biopharm Rhône Ltd, Glasgow, UK.

OTA Quantification in Cocoa Beans

The dried cocoa bean samples were frozen at -80°C, then ground. Fifty grams of ground beans were extracted in 200 ml of solvent (acetonitrile/water, 60/40, v/v). Four millilitres of filtered extract were diluted in 44 mL of phosphate buffer. The mixture was purified on an immunoaffinity column (Ochraprep, Rhône Diagnostics, Scotland). OTA was eluted by 3 ml of methanol and evaporated till dry in a nitrogen stream at 70°C. The residue was resuspended in 1 ml of the mobile phase (water/acetonitrile/acetic acid, 51:48:1, v/v).

Quantification was by HPLC using the previously described method.

RESULTS

Fungal Microflora from Cocoa Beans

Identification of the total fungal microflora in samples from different cocoa fermentation is given in Table 1. Independent of the type of post-harvest, cocoa beans are contaminated by moulds. This contamination is greater when the pods were damaged. The strains mainly belong to the genera *Penicillium, Aspergillus, Scopulariopsis, Syncephalastrum, Mucor, Geotrichum, Trichoderma, Rhizopus, Fusarium*. Figure 1 shows the phenotypic appearance of some isolated fungi. Pod integrity, and in a lesser degree, the deadline of pod – opening affected quality and quantity diversity of isolated moulds. The genus *Aspergillus* represents several different species which are black and correspond to the section Nigri, which contains species known to produce OTA (*Aspergillus niger agg* and *Aspergillus carbonarius*) (Amezqueta et al. 2008; Mounjouenpou et al., 2008). Wounded pods had high proliferation of *A. carbonarius, Fusarium spp* and *A. niger* in the pod openings.

Moulds were found in all types of post-harvest treatment. Their number varied depending on the sampling stage. When the pod is undamaged, the mould content of beans was not detectable before fermentation, i.e. less than 10 $CFU.g^{-1}$. With fermentation and sun drying, the content was increased considerably to the value of 4.7 ± 0.6 x 106 $CFU.g^{-1}$. Even when pod was damaged and beans processed immediately, mould content was still less important.

The highest level of contamination was obtained with beans from damaged pods and deferred pod-opening. The content obtained was 11.7 ± 0.9 x 106 $CFU.g^{-1}$. After fermentation and sun drying, this content was decreased to a value of 3.7 ± 0.5 x 106 $CFU.g^{-1}$.

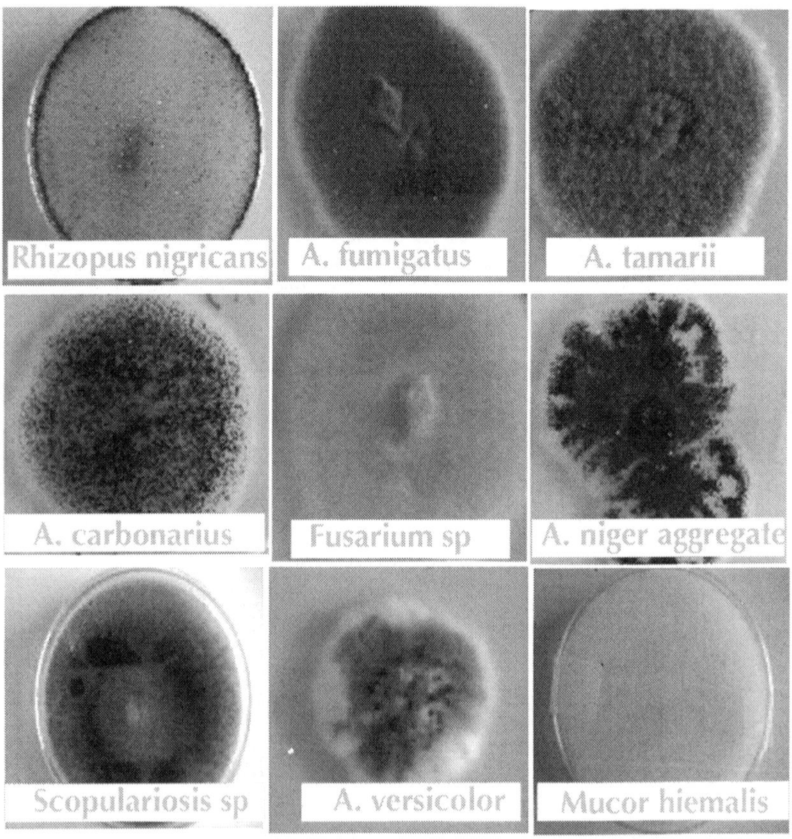

Figure 1: Phenotypic aspect of some isolated moulds.

Table 1: Identification of the total fungal microflora in samples from different cocoa fermentation

Post-harvest conditions	Sampling stage	Total moulds (CFU.g^{-1})	Isolated strains
Immediate fermentation with undamaged pods	Before fermentation	nd	-
	After fermentation and sun drying	4.7 ± 0.6 x 10^{6}	A. tamarii,A. fumigatus, Rhizopus *nigricans, A. niger* agg
Immediate fermentation with damaged pods	Before fermentation	0.9 ± 0.1 x 10^{6}	*A. niger* **agg,** *Fusarium spp, A. fumigatus, Geotrichum,*
	After fermentation and sun drying	3.0 ± 0.2 x 10^{6}	*Rhizopus nigricans, A. niger* agg; *A. flavus, Trichoderma virens*
Delayed fermentation with	Before fermentation	11.7 ± 0.9 x 10^{6}	*Scopulariopsis spp, A. niger* agg, *Syncephalastrum racemosum, Geotrichum spp, A. fumigatus, Rhizopus nigricans, Mucor spp, P. crustosum, A. carbonarius, Fusarium* spp
damaged pods	After fermentation and sun drying	3.7 ± 0.5 x 10^{6}	*P. crustosum, P. sclerotiorum, Fusarium spp, Scopulariopsis spp, Rhizopus nigricans, A. flavus, Trichoderma virens, A. niger* agg

nd: <10 CFU.g^{-1}

OTA on Cocoa and Rice Medium from Producing Moulds Isolated from Cocoa Beans

The ability to produce OTA by isolated *Aspergillus carbonarius* and *Aspergillus niger agg* were studied using official medium (rice medium) and cocoa medium.

All strains of *Aspergillus carbonarius* had a high ochratoxinogenic activity which varied depending on the culture medium. OTA production was greater on rice medium with a content of 573.4 to 2772.0 $ng.g^{-1}$ after twenty-one days of culture. On cocoa medium, values of 50.6 to 110.7 $ng.g^{-1}$ was obtained (Table 2).

Compared to *Aspergillus carbonarius*, strains of *Aspergillus niger agg* were less toxinogenic whatever the type of culture medium. OTA production ranged from undetectable (<0.03ng.g^{-1}) to a maximum content of 0.20 $ng.g^{-1}$ on cocoa medium. On rice culture medium, the content was from undetectable to 3.6 $ng.g^{-1}$.

Table 2: OTA production by moulds

Strains	OTA production (ng.1[1])	
	Cocoa medium	Rice medium (FDA)
A. carbonarius 1	50.6 ± 0.3	573.4 ± 1.6
A. carbonarius 2	110.7 ± 0.6	2772.0 ± 1.7
A. niger 1	nd	nd
A. niger 2	0.20 ± 0.01	3.5 ± 0.0
A. niger 3	**0.05** ± 0.02	3.6 ± 0.0

OTA Quantification after Cocoa Fermentation and Sun Drying

Table 3 showed the OTA level, and toxinogenic microflora associated with cocoa beans resulting from different post harvest treatments.

When the pod was undamaged, the toxinogenic microflora associated with cocoa bean was mostly consisted of *A. niger agg*. The OTA content was in this case between nd (not detectable: <0.03 $ng.g^{-1}$) and 0.03 $ng.g^{-1}$, which remained below 2 $ng.g^{-1}$ (the European limit defined for cocoa beans).

When pods are damaged, a maximum level of 12.14 $ng.g^{-1}$ was observed prior to fermentation: this content was much higher than the tolerable doses. Toxigenic species associated with these beans are *Aspergillus carbonarius* and *Aspergillus niger agg*.

With fermentation and sun drying, there was a decrease of the OTA content of beans to a value of 1.01 $ng.g^{-1}$. Associated ochratoxigenic microflora was only *A. niger agg.*

Table 3: OTA level on cocoa beans and associated toxigenic moulds

Post-harvest conditions	Sampling stage	Isolated ochratoxigenic strains	**OTA in cocoa** beans **($ngeg^{-1}$)**
Immediate fermentation with undamaged pods	Before fermentation	-	nd*
	After fermentation and sun drying	*A. niger* agg	0.03 ± 0.00
Immediate fermentation with damaged pods	Before fermentation	*A. niger* agg	0.83 ± 0.02
	After fermentation and sun drying	*A. niger* agg	1.33 ± 0.03
Delayed fermentation with damaged pods	Before fermentation	*A. niger* agg, *A. carbonarius,*	12.14± 0.10
	After fermentation and sun drying	*A. niger* agg	1.01 ± 0.02

*: < 0.03 $ng.g^{-1}$

DISCUSSION

This study was conducted in the region of Kumba in Cameroon. The climate is equatorial. It rains throughout the year and rainfall can reach 3000 mm / year. This rainfall implies high humidity (~ 90%), which promotes growth of many microorganisms that can influence the final quality of the cacao from that region.

Irrespective of pod condition and post harvest conditions, a large increase in fungal flora was found after fermentation and sun drying (qualitatively and quantitatively). The main moulds isolated in our study belonged to the genera *Penicillium, Aspergillus, Mucor, Scopulariopsis, Syncephalastrum, Geotricum, Trichoderma, Rhizopus,* Fusarium with some species known to produce OTA (Aspergillus niger, *Aspergillus carbonarius*). Similar moulds were founded by Mounjouenpou et al (2008) when studying the filamentous fungi during cocoa processing in Cameroon. Moulds belonging mainly to the species Rhizopus stolonifer (Ehrenb.) Lind., *A. niger* aggregate, *Aspergillus* flavus Link,

Penicillium citrinum Thom and A. carbonarius to a minor extent were isolated from stored cocoa beans by Amezqueta et al. (2008).

Our results differed from those quoted in the literature for the fungal microflora associated with fermented beans (Maravalhas, 1966) or dried beans (Buting, 1928; Dade, 1928; Ciferri, 1931). In those publications, *Aspergillus fumigatus, Aspergillus glaucus, Mucor spp* and *Penicillium spp* were isolated.

During cocoa fermentation, there was competition between all the species present. Those with high-speed growth (*Mucor, Rhizopus* spp) were able to colonize first the medium before *Aspergillus* and *Penicillium genera*. Generally, drying contributed to reduce microflora.

During the solar drying, sensitive species disappeared in favor to soil moulds, and more generally, environmental species. *Aspergillus niger agg* was found in all conditions: from fresh beans to fermented and dried beans. Contamination by *Aspergillus carbonarius* was mostly found in beans from the damaged pods and deferred pod - opening. This high contamination may come from the fact that when the pods were middle-open, cocoa beans were in direct contact with air and soil, possible source of A. carbonarius. When the pods are whole, strains of A. carbonarius could be less competitive to settle in the milieu. This situation would change when the pods are damaged.

Our results showed that *Aspergillus carbonarius* (100% of strains) were able to produce OTA in cocoa. Strains of *Aspergillus carbonarius* were more toxigenic than those of *A. niger agg*. As in grapes (Belli et al., 2006), *Aspergillus carbonarius* is the main producing strain of OTA in cocoa. This coincide with that of Sage et al. (2002), Bellı´ et al. (2005), Leong et al. (2006), and Astoreca et al. (2007). Despite OTA being a stable metabolite, the mycotoxin content decreased with incubation time. Other studies (Bellı´ et al. 2004; Esteban et al. 2006; Astoreca et al. 2007; Romero et al. 2007) have also noted this phenomenon.

Some strains of A. niger agg (66.6%) were able to produce OTA in cocoa. The proportion of OTA-producing strains of A. niger agg varied depending on the context: Taniwaki et al. (2003) have shown that on coffee, 75% of Aspergillus ochraceus and 3% of A. niger agg produce OTA. On cocoa, Amèzqueta et al. (2008) were found any isolated strains of A. niger agg was ochratoxinogenic. Other authors (Taniwaki et al. 1999, 2003; Sùarez-Quiroz et al. 2004; Illic et al. 2007; Leong et al. 2007) have reported that 1–9% of *Aspergillus niger* strains isolated from coffee beans produce the toxin. This difference could be attributed to a natural selection in the strain or to adverse environmental conditions.

REFERENCES

1. Abarca M. L., Bragulat M. R., Castella G., & Cabanes F. J. (1994). Ochratoxin A production by strains of *Aspergillus niger var. niger*. Applied and Environmental Microbiology, 60, 2650-2652.
2. Abrunhosa, L., Paterson, R. R. M., Kozakiewicz, Z., Lima, N., & Venancio, A. (2001). Mycotoxin production from fungi isolated from grapes. Letters in Applied Microbiology, 32, 240-242. http://dx.doi.org/10.1046/j.1472-765X.2001.00897.x
3. Amézqueta, S., González-Peñas, E., Dachoupakan, C., Murillo-Arbizu, M., López de Cerain, A., & Guiraud J. P. (2008). OTA-producing fungi isolated from stored cocoa beans. Letters in Applied microbiology, 47, 197-201. http://dx.doi.org/10.1111/j.1472-765X.2008.02409.x
4. Astoreca, A., Magnoli, C., Barberis, C., Chiacchiera, S.M., Combina, M., & Dalcero, A. (2007). Ochratoxin A production in relation to ecophysiological factors by *Aspergillus* section Nigri strains isolated from different substrates in Argentina. Sci Total Environ, 388, 16–23. http://dx.doi.org/10.1016/j.scitotenv.2007.07.020
5. Belli, N., Bau, M., Marin, S., Abarca, M. L., Ramos, A. J., & Bragulat, M. R. (2006). Mycobiota and ochratoxin A fungi from Spanish wine grapes. Int. J. of Food Microbiology, 111, S40-S45. http://dx.doi.org/10.1016/j.ijfoodmicro.2006.03.011
6. Bellı´, N., Ramos, A. J., Sanchı´s, V., & Marı´n, S. (2004). Incubation time and water activity effects on ochratoxin A production by *Aspergillus* section Nigri strains isolated from grapes. Lett Appl Microbiol, 38, 72–77. http://dx.doi.org/10.1046/j.1472-765X.2003.01445.x
7. Bellı´, N., Ramos, A.J., Coronas, I., Sanchı´s, V., & Marı´n, S. (2005). *Aspergillus* carbonarius growth and ochratoxin A production on a synthetic grape medium in relation to environmental factors. J

Appl Microbiol, 98, 839–844. http://dx.doi.org/10.1111/j.1365-2672.2004.02469.x

8. Blanc, M., Pittet, R., & Munoz-Box, R. (1998). Behavior of ochratoxin A during green coffee roasting and soluble coffee manufacture. J. of Agri.and Food chemistry, 46, 673-675.
9. Buting, R. H. (1928). Fungi occurring in cocoa beans. Department of Agriculture Gold Coast year Book bull, 16, 44-62. http://dx.doi.org/10.1021/jf9707703
10. Ciferri, R. (1931). Studies on cocoa. Journal of the Department of Agriculture Porto Rico, 15, 223–286.
11. Dade, H. A. (1928). Internal moulding of prepared cocoa. Department of Agriculture Gold Coast Year Book bull, 16, 74–100.
12. Durand, N. (2009). Study on ochratoxin A contamination of coffee batches in the Kenyan context, in relation to cultivation methods and post harvest processing treatments. Food Add. and Contaminants.
13. Duris, D., Mburu, J. K. Durand, N., Clarke, R., John, M., & Guyot, B. (2010). Ochratoxin A contamination of coffee batches from Kenya in relation to cultivation methods and post-harvest processing treatments. Food Additives and Contaminants Part A – Chemistry Analysis Control Exposure & Risk Assessment, 27, 836-841.
14. Esteban, A., Abarca, M. L., Bragulat, M. R., & Caban˜es, F. J. (2006) Study of the effect of water activity and temperature on ochratoxin A production by *Aspergillus* carbonarius. Food Microbiol, 23, 634–640. http://dx.doi.org/10.1016/j.fm.2005.12.006
15. Illic, Z., Bui, T., Tran-Dinh, N., Dang, M. H. V., Kennedy, I., & Carter, D. (2007). Survey of Vietnamese coffee beans for the presence of ochratoxigenic Aspergilli. Mycopathologia, 163, 177–182. http://dx.doi.org/10.1007/s11046-007-0099-0
16. Gareis, M., & Scheuer, R. (2000). Ochratoxin A in meat and meat products. Archiv für Lebensmittelhygiene, 51, 102–104.
17. Hocking, A. D. (1991). Isolation and identification of xerophilic fungi in stored commodities. In fungi and Mycotoxines in Stored Products. ACIAR Proceeding n° 36, Bangkok, Thailand. 23-26 April pp. 65-72.

18. Höhler, D. (1998). Ochratoxin A in food and feed: occurrence, legislation and mode of action. Zeitschrift für Ernährungswissenschaft, 37, 2–12.
19. Hurst, M. J., & Martin, R. A. (1998). High-performance liquid chromatography determination of Ochratoxin A in artificially contaminated cocoa beans using automated clean-up. J. of Chromatography A, 810, 89–94. http://dx.doi.org/10.1016/S0021-9673(98)00202-7
20. Jorgensen, K. (1998). Survey of pork, poultry, coffee, beer and pulses for Ochratoxin A. Food Add. and Conta, 15, 550–554. http://dx.doi.org/10.1080/02652039809374680
21. Kuiper-Goodman, T. (1996). Risk assessment of ochratoxin A: an update. Food Add. And Conta., 13, 535–557.
22. Kuiper-Goodman, T., Scott, & P. M. (1989). Risk assessment of the mycotoxin OchratoxinA. Biometrics Biomedical and Environmental Sciences, 2, 179–248.
23. Leong, S. L., Hocking, A. D., & Scott, E. S. (2006). Effect of temperature and water activity on growth and ochratoxin A production by Australian *Aspergillus carbonarius* and A. niger isolates on a simulated grape juice medium. Int J Food Microbiol, 110, 209–216. http://dx.doi.org/10.1016/j.ijfoodmicro.2006.04.005
24. Leong, S. L., Hien, L. T., An, T. V., Trang, N. T., Hocking, A. D., & Scott, E. S. (2007) Ochratoxin A-producing Aspergilli in Vietnamese green coffee beans. Lett Appl Microbiol, 45, 301–307. http://dx.doi.org/10.1111/j.1472-765X.2007.02189.x
25. Mantle, P. G., McHugh, & K. M. (1993). Nephrotoxic fungi in foods from nephropathy households in Bulgaria. Mycological Research, 97, 205–212. http://dx.doi.org/10.1016/S0953-7562(09)80242-6
26. Maravalhas, M. (1966). Mycological deterioration of cocoa beans during fermentation and storage in Bahia. Int.Chocolate Review, 21, 375–378.
27. Mounjouenpou, P., Gueule, D., Guyot, B., Tondje, P. R., Fontana-Tachon, A., & Guiraud, J. P, (2008). Moulds producing ochratoxin A during cocoa processing in Cameroon. International Journal of

Food Microbiology, 121, 234 – 241. http://dx.doi.org/10.1016/j.ijfoodmicro.2007.11.017

28. Mounjouenpou, P., Gueule, D., Ntoupka, M., Durand, N., Fontana-Tachon, A., Guyot, B., & Pierre Guiraud, J. P., (2011). Influence of post-harvest processing on ochratoxin A (OTA) content in cocoa and on consumer exposure in Cameroon, World Mycotoxin Journal, 2, 141-146. http://dx.doi.org/10.3920/WMJ2010.1255
29. Nakajima, M., Tsubouchi, H., Miyabe, M., & Ueno, Y. (1997). Survey of aflatoxin B1 and Ochratoxin A in commercial green coffee beans by high- performance liquid chromatography linked with immunoaffinity chromatography. Food and Agri. Immunology, 9, 77–83. http://dx.doi.org/10.1080/09540109709354938
30. O'Callaghan, J., Caddick, M. X., & Dobson, D. W. (2003). A polyketide synthase gene required for Ochratoxin A biosynthesis in *Aspergillus* ochraceus. Microbiology, 149, 3485–3491. http://dx.doi.org/10.1099/mic.0.26619-0
31. Pitt J. I., Basilico J. C., Abarca M. L., & Lopez C. (2000). Mycotoxins and toxinogenic fungi. Medical Mycology, 38, 41-46.
32. Pittet, A., Tounare, D., Huggett, A., & Viani, R., (1996). Liquid chromatographic determination of ochratoxin A in pure and adulterated soluble coffee using an immunoaffinity column cleanup procedure. J. of agri. and food chemistry, 44, 3564–3569. http://dx.doi.org/10.1021/jf9602939
33. Romero, S. M., Patriarca, A., Ferna´ndez-Pinto, V., & Vaamonde, G. (2007). Effect of water activity and temperature on growth of ochratoxigenic strains of *Aspergillus* carbonarius isolated from Argentinean dried vine fruits. Int J Food Microbiol, 115, 140–143. http://dx.doi.org/10.1016/j.ijfoodmicro.2006.10.014
34. Sage, L., Krivobok, S., Delbos, E., Seigle-Murandi, F., & Creppy, E. E. (2002). Fungal flora and ochratoxin A production in grapes and musts from France. J Agric Food Chem, 50, 1306–1311. http://dx.doi.org/10.1021/jf011015z
35. Samson, R. A., Hoekstra, E. S., Frisvad, J. C., & Filtenborg, O. (1995). Introduction to food-borne fungi. Centraalbureau voor Schimmelcultures, Baarn, Delft Netherlands 7th edition.

36. Skaug, M. A. (1999). Analysis of Norwegian milk and infant formulas for Ochratoxin A. Food Add. and Conta., 16, 75–78. http://dx.doi.org/10.1080/026520399284235

37. Suàrez-Quiroz, M., Gonzàlez-Rios, O., Barel, M., Guyot, B., Schorr-Galindo, S., Guiraud, & J. P. (2004). Study of Ochratoxin A-producing strains in coffee processing. Journal of Food Science and Technology, 39, 501–507.

38. Suàrez-Quiroz, M., Gonzàlez-Rios, O., Barel, M., Guyot, B., Schorr-Galindo, S., & Guiraud, J. P. (2005). Effect of post-harvest processing procedure on OTA occurrence in artificially contaminated coffee. International Journal of Food Microbiology, 103, 339-345. http://dx.doi.org/10.1016/j.ijfoodmicro.2004.11.044

39. Taniwaki, M. H., Pitt, J. I., Urbano, G. R., Teixeira, A. A., & Leita˜o, M. F. F. (1999). Fungi producing OTA in coffee. 18th Int Scient Coll Coffee, Helsinki, pp. 239–247.

40. Taniwaki, M. H., Pitt, J. I., Texeira, & Iamanaka B. T. (2003). The source of ochratoxin A in Brazilian coffee and its formation in relation to processing methods. International J. of Food Microbiol., 82, 173-179. http://dx.doi.org/10.1016/S0168-1605(02)00310-0

41. Thirumala-Devi, K., Mayo, M. A., Reddy, G., Emmanuel, K.E., Larondelle, Y., & Reddy, D. V. R., (2001). Occurrence of Ochratoxin A in black pepper, coriander, ginger and turmeric in India. Food Add. and Conta., 18, 830–835.

42. Tournas, V., Stack, M. E., Mislivec, P.V., Koch, H. A., & Bandler, R. (2001). Bacteriological Analytical Manual Online. US Food & Drug Administration (http//www.cfsan.fda.gov/~ebam/bam-18.html).

43. WHO (World Health Organization). (1996). Evaluation of certain food additives and contaminants. 44th Report of JECFA; WHO Technical Report series 859; WHO: Geneva, Switzerland.

Chapter 6

Traditional Procedures of Cocoa Processing and Occurrence of Ochratoxin - A in the Derived Products

Pauline Mounjouenpou[1], Joseph Amang A Mbang[1], Bernard Guyot[2], Joseph-Pierre Guiraud[2]

[1]Institut de Recherche Agricole pour le Développement (IRAD), BP 2067, Yaoundé, Cameroun

[2]UMR Qualisud (CIRAD, Université Montpellier II, UM1, Sup Agro), place E. Bataillon, 34095 Montpellier Cedex 5, France

ABSTRACT

Cocoa beans from several types of post-harvest treatments were processed into cocoa butter, cocoa powder and chocolate spread using traditional processing techniques. OTA was analyzed in two hundred and sixteen (216) samples of cocoa and cocoa derived products using

HPLC with fluorimetric detection. It was found that OTA content in cocoa nibs depended only on the state of cocoa pod and the time duration of cocoa pod-opening. Cocoa beans from damaged cocoa pods and pod-opening delayed by 10 days were the most contaminated by OTA. Neither roasting using fire wood, nor shelling by pounding in a mortar could reduce significantly the OTA content in cocoa beans. Apart from cocoa butter, the other cocoa derived products also had high OTA contents. Sixty seven (67%) of chocolate spread and cocoa powder samples were contaminated by OTA, and 50 % had contents superior to the limit defined by the European Union. Because of the fragile pellicle of cocoa beans, due to damaged pods and cocoa pod-opening delayed by 10 days, the contamination by OTA was no longer superficial and shelling does not longer contribute to reduce significantly the OTA content in beans. Like with modern procedures, traditional techniques of cocoa processing do not have significant influence on OTA content of derived products. Only the raw material (quality and state of cocoa pod) can influence the OTA content of cocoa derived products.

INTRODUCTION

Ochratoxin A (OTA) is a mycotoxin produced mainly by fungi of *Aspergillus* and *Penicillium* genera. It is mainly produced by *Aspergillus carbonarius, Aspergillus niger, Aspergillus ochraceus* and A. westerdijkiae in tropical zones and by *Penicillium verrucosum* or P. *nordicum* in temperate zones [1, 36, 40]. In cocoa, only black Aspergillus of Nigri section (A. *carbonariu*s and A. niger agg) are found [27]. The *Ochratoxin* - A has *nephrotoxic*[25], *immunotoxic, teratogenic* and *carcinogenic* effects [19, 21, 22]. Following research results obtained on animals, the International Agency for Research on Cancer (IARC) classified OTA as carcinogenic in humans (group 2B). OTA occurs naturally in a wide variety of raw commodities and finished food products of plant origin such as cereals, coffee beans, cocoa beans, pulses, and dried fruits as well as cocoa derived products. *Ochratoxin* A presence has also been detected in lower amounts in meat, eggs and milk as a product of the carryover of animals feed with contaminated feedstuffs [2, 9, 12, 20, 29, 30, 32, 33, 37, 38, 39]. occurrence of OTA in cocoa powder and cocoa products was reported since 2000 by Engel [15]. During this decade, incidence and intake trends remained basically constant [32, 35].

The presence of OTA in cocoa is mainly related to cocoa bean shells and fat-free cocoa solids (cocoa powder) [3, 18]. Cocoa production in Cameroon is estimated at 200 000 tons per year. This production is from about 500 000 farmers and is distributed in three large zones of production. In Cameroon, the government does not control cocoa commercialization. Several actors intervene in cocoa commercialization; the middlemen buy at a derisory price (USD 1.6 per kg of cocoa) from farmers to resell at a higher price to exporters (about USD 3 per kg). In this case, farmers are the losers. National statistics have revealed a high level of poverty in the rural areas of Cameroon.

Therefore, to improve the livelihood of farmers' particularly that of cocoa-farmers, the Government of Cameroon has adopted an agricultural policy that incites local consumption of cocoa. Following this, easy and applicable technologies were developed and disseminated to farmers, with the idea to process cocoa using local materials available in farmers' area. Although several authors [9, 18] have studied the occurrence and OTA content of cocoa derived products obtained from modern and improved technics of cocoa processing, it is not yet the case of those resulting from the traditional technics. The goal of this work was to assess the effects of traditional technics of cocoa processing on OTA content of some derived products of cocoa such as cocoa butter, cocoa powder, and chocolate spread.

EXPERIMENTAL SECTION

Cocoa

Cocoa (Theobroma cacao L.) was harvested in Kumba (Cameroon), during 2005, 2006 and 2007 cocoa seasons. Kumba, of the South West Region, is the biggest cocoa producing zone in Cameroon: accounting for more than 60% of the national production, which stands at 180 000 tons annually. The type of cocoa grown in this region is the Amazonian. The climate is equatorial type with a total rainfall of 3000 mm year^{1}

Chemicals

Liquid chromatography grade methanol, acetonitrile and acetic acid were purchased from Carlo Erba (Carlo Erba Reagenti, Milan, Italy). Water was purified by distillation and passage through a Milli-Q system (Millipore, Bedford, MA, USA). Phosphate Buffer Solution (PBS) was used for the dilution step. Ochratoxin A certified standard solution (10.15 - 0.14 mg/mL) was purchased from Biopure (Tulln, Austria). For clean-up step OCHRAPREP immunoaffinity column (IAC) was used (R-Biopharm Rhône Ltd, Glasgow, UK).

Post-harvest treatments of cocoa and sampling

According to the type of fermentation (i.e. heap or box), the physical appearance of the cocoa pod appearance (i.e. damaged or not) or schedule for the splitting of the cocoa pods (immediately or delayed of 10 days), cocoa nibs went through four different types of post-harvest treatments (table 1).

Table 1: Different post-harvest treatments

Treatment	Cocoa pod appearance	Pod opening	Type of fermentation
T1	Intact	immediate	box
T2	Intentionally damaged	After 10 days	Heap
T3	Damaged on the tree	immediate	Heap
T4	Damaged on the tree	After 10 days	Heap

Fermented sun-dried cocoa from different technological treatments are used to fabricate derived products which are cocoa butter, cocoa powder and chocolate spread.

BUTTER EXTRACTION

Cocoa beans were placed in a locally made aluminum pot and roasted by stirring regularly for about 60 minutes under about 50°C.

The roasted cocoa beans were introduced into a mortar and pounded lightly to facilitate removal of the husk by winnowing. Thereafter, the beans were finely ground into a paste with the aid of a mill. The paste obtained was mixed with water and cooked for several hours to extract the cocoa butter. The extracted cocoa butter was collected, washed using distilled water and filtered using a clean tissue.

Fabrication of Chocolate Spread and Cocoa Powder

After extracting the butter, cocoa paste was obtained by evaporating the water. The cocoa paste with reduce lipid contents was collected and mixed with sugar and refined oil to obtain chocolate spread which can be stored for about 3 months. The paste of cocoa was dried to obtain cocoa cake. The dry cake was finely ground until the desired grade. Cocoa powder obtained can be consumed as hot drink with sugar during breakfast or as cold chocolate drink.

Sampling

Sampling was done at different steps of cocoa processing: fermented sun-dried cocoa (A), roasted cocoa (B), nibs (C), cocoa butter (D), cocoa powder (E) and chocolate (F). A total of seventy two (72) samples were analyzed during each cocoa season, given a total of 216 samples for the entire study period.

Titration of OTA

The dried cocoa bean samples were frozen at -80°C then ground. Fifty grams of ground beans or cocoa powder were extracted in 200 mL of solvent (acetonitrile/water, 60/40, v/v). Four millilitres of filtered extract were diluted in 44 mL of phosphate buffer. The mixture was purified on an immunoaffinity column (Ochraprep, R-Biopharm, Glasgow, UK). OTA was eluted by 3 mL of methanol and evaporated till dry in a nitrogen stream at 70°C. The residue was resuspended in 1 mL of mobile phase (water/acetonitrile/acetic acid, 51:48:1, v/v/v). OTA was quantified on the extracts by HPLC with fluorimetric detection (Shimadzu LC-10 ADVP, Japan) [34]. The operating conditions were as follows: 100 μL injection loop, C18 reverse phase HPLC column, ODS particle size of 5 μm with an identical pre-column, thermostatically controlled at 35°C, isocratic flow of 1 mL.min-1, an excitation wavelength of 333 nm and an emission wavelength of 460 nm. Contents were calculated from a calibration curve established from a standard (1 $\mu g.mL^{-1}$; ref PD 226 R-Biopharm, Glasgow, UK).

RESULTS AND DISCUSSION

Content of OTA in Beans according to the Type of Post-Harvest Treatment

Table 2 presents OTA contents of different fermented sun-dried cocoa according to the type of post-harvest treatment. OTA contents were ranged between detection limit (dl < 0.03 $ng.g^{-1}$) and 0.15 $ng.g^{-1}$ when beans were provided from not damaged cocoa pods T1; which remains less than 2 $ng.g^{-1}$ which is the tolerable set defined for cocoa beans (European Commission, 2006). Low OTA content of cocoa was also obtained in beans from cocoa pods damaged on the tree, and pod opening immediately (T3). However, when pods were not intact and cocoa pod - opening was deferred (T2, T4), cocoa beans resulting from those treatments were the most contaminated by OTA (table 2). Maximum content of 20.75 $ng.g^{-1}$ (value which is higher than tolerable set) was observed with treatment T4. Therefore, the contamination by OTA did not depend on the type of fermentation (box or heap). The presence of OTA during cocoa post-harvest depended on the state of cocoa pod and the deadline of cocoa pod - opening. These results are in accordance with those of Bastide and al. [6] who showed that the presence of OTA during post-harvest treatment of cocoa depends on the cocoa pod state. According to the same authors, bean OTA content depended on the phytosanitary condition of pods, with 4.2 $ng.g^{-1}$ for pods damaged by insects, 19.8 ng/g for wounded pods, 7.2 $ng.g^{-1}$ for rotten cocoa pods and 3.4 $ng.g^{-1}$ for mummified pods. Studies made by *Suàrez-Quiroz* and al., [42] have shown that it was possible to compare high contents in OTA to the presence of *d'Aspergillus carbonarius*. Contamination by *Aspergillus carbonarius* was mainly found in damaged cocoa pods with deferred cocoa pod - opening.

When cocoa pods were partially covered, cocoa beans were in direct contact with air and soil, which led to high *ochratoxinogenics* activities [28].

OTA Content of Beans According to Cocoa Season Year

Our results have shown that, treatments T2 and T4 are generally those that lead to high contamination of cocoa by OTA. It was necessary to observe if the year of cocoa season could have an effect on this contamination. OTA contents vary. Table 3 showed that irrespectively of the cocoa season, the maximum contamination was always obtained with beans from tree damaged cocoa pods and cocoa pod - opening delayed by 10 days (T4). Only the maximum content varies according to year. The 2005 cocoa season was the most polluted with a content of 20.75 $ng.g^{-1}$. Only half (10 $ng.g^{-1}$) of this content was obtained in 2006 and 2007 cocoa seasons. This study was done in Kumba area which is an agro-ecological zone with a monomodal rainfall. The mean annual rainfall is 3000 mm. This implicates a high humidity (90 %). The difference in OTA content obtained could be due to other environmental factors that are difficult to control such as the level of air pollution by ochratoxinogenic fungi, speed of wind, etc.

Bastide and al., [6] have also shown that OTA contents could vary with the cocoa season.

Table 2: OTA contents of cocoa beans and different cocoa derived products

Type of treatment	Sampling	OTA content: $ng.g^{-1}$ (2005 season)	OTA content: $ng.g^{-1}$ (2006 season)	OTA content: $ng.g^{-1}$ (2007 season)
IT1	Cocoa beans	0.15 ± 0.01	nd	0.6 ± 0.00
	Roasted Cocoa	nd	0.03 ± 0.01	nd
	Nibs	nd	0.02 ± 0.00	0.01 ± 0.00
	butter	nd	nd	nd
	Cocoa powder	0.04 ± 0.01	nd	0.03 ± 0.02
	Chocolate spread	nd	0.01 ±0.01	0.03 ± 0.00

T2	Cocoa beans	0.16 ± 0.02	1,14 ± 034	5.34 ± 0.21
	Roasted Cocoa	0.08 ± 0.00	0.92 ± 0.02	5.12 ± 0.82
	Nibs	nd	nd	4.08 ± 0.11
	butter	0.01 ± 0.00	nd	0.02 ± 0.00
	Cocoa powder	nd	0.01 ± 0.01	1.55 ± 0.07
	Chocolate spread	nd	nd	1.95 ± 0.36
T3	Cocoa beans	0.04 ± 0.00	0.25 ± 0.01	0.05 ± 0.00
	Roasted Cocoa	0.05 ± 0.04	0.5 ± 0.21	0.03 ± 0.03
	Nibs	nd	nd	0.02 ± 0.02
	butter	0.02 ± 0.00	0.01 ±0.01	0.01 ±0.01
	Cocoa powder	0.03 ± 0.01	0.13 ± 0.00	nd
	Chocolate spread	nd	0.03 ± 0.02	nd
T4	Cocoa beans	20.75 ± 1/2	10.0 ± 0.22	10.0 ± 1.22
	Roasted Cocoa	19.86 ± 0.89	10.11 ± 1.04	9.80 ± 1.01
	Nibs	18.19 ± 1.07	9.45 ± 0.99	8/2 ± 1.33
	butter	1.88 ± 032	1.10 ± 0.44	0.93 ± 0.61
	Cocoa powder	16.15 ± 1.11	7/2 ± 0.77	6.99 ± 033
	Chocolate spread	17.00 ± 1.58	8.89 ± 0.69	7.33 ± 1.01

Table 3: Influence of the cocoa season year on beans OTA content

Type of treatment	OTA content: ng g^{-1}		
	2005 season	2006 season	2007 season
T2	0.16 ± 0.02	1.14 ± 0.34	5.34 ± 0.21
T4	20.75 ± 1.22	10.0 ± 0.22	10.0 ± 1.22

TECHNOLOGICAL TREATMENT OF COCOA BEANS AND OTA CONTENT

OTA content in roasted beans

The OTA content of roasted beans from undamaged cocoa pods ranged between the nd and 0.03 ± 0.01 $ng.g^{-1}$ (value inferior to the tolerable limit for cocoa beans). In addition, significant differences in OTA content were not observed with roasted beans obtained from pods that were damaged by birds or cutlasses during harvesting. However, late cocoa pod-opening (delayed by 10 days) of the same beans contributed to obtain fermented sun-dried cocoa with maximum content of OTA 5.34 ± 0.21$ng.g^{-1}$ for T2 and 20.75 ± 1.22 $ng.g^{-1}$ for T4. After roasting using fire wood for almost an hour, OTA contents are 5.12 ± 0.82 $ng.g^{-1}$ and 19.86 ± 0.89 $ng.g^{-1}$ respectively. Therefore, roasting did not have an impact on beans OTA content. OTA conserves its stability during most of the steps of foods processing such as cooking, washing and fermentation [5]. Boudra and al. [8] have shown that OTA is resistant to heat and a maximum of 20% of OTA in wheat is decomposed by dry heat at 100 °C for 160 minutes or at 150 °C for 32 minutes. Suàrez-Quiroz and al., [43] observed a reduction of OTA content in coffee after thermal treatment. These authors explained the reduction by a possible complexation with other molecules of the milieu. Mounjouenpou and al. [26] also noted a reduction of OTA content with heating. This reduction depends on the pH of the milieu with the largest reduction observed at pH>7 (alkaline medium) and at 85 °C. During cocoa roasting, the final temperature of beans reaches 100 to120°C for a period of 15 to 70 minutes [4, 23, 31].In these conditions, it is not expected that roasting reduces OTA at higher levels because a significant degradation can only be possible at 180-200°C [16, 43]. Traditional procedures of cocoa beans roasting applied in rural zones that consist to roast in a pot does not therefore present any impact on OTA content of the product because the temperature of beans remains less than 100°C.

OTA CONTENT OF NIBS

Roasted cocoa beans was shelled by pounding in a mortar and winnowing to obtain the nibs. When pod-opening was immediate, (T1 and T3), OTA content in nibs remained less than the limit of detection (<0.03 ng.g^{-1}). However, when nibs result from a treatment where cocoa pod is damaged, and pod-opening was deferred, almost all the OTA of roasted beans was found in nibs. When cocoa pods are intentionally damaged, OTA passes from 5.12 ± 0.82 ng.g^{-1} in roasted beans to 4.08 ± 0.11 ng.g^{-1} in nibs during the 2007 season. For cocoa pods naturally injured, values are respectively 19.86 ± 0.89 ng.g^{-1} and 18.19 ± 1.07 ng.g^{-1} in 2005. Generally, literature reveals a reduction of OTA content in cocoa nibs due to shelling. Gilmour, and Lindblom [18] analyzed OTA in 15 samples pairs of cocoa shell and bean. Results showed that a mean of 48 % (ranging from 25 to 72%) of OTA present in cocoa bean was eliminated together with the shell. In a study were cocoa shell were eliminated manually, Amèzqueta and al., [43] observed a reduction of OTA content (>95 %) in 14 out of 22 samples, of 65 to 95 % in 6 out of 22 samples and of less than 50 % in only one of the samples. Contrary to the literature, our results showed that shelling does not necessary reduce OTA content in cocoa nibs. With damaged cocoa pods and deferred pod-opening of 10 days, cocoa beans start to germinate. The protector pellicle of almonds is no longer entire as it breaks out in order to allow the young plant grow. With fermentation and drying, beans are more fragile. It crumbs with a least friction between fingers. Ochratoxinogenic fungi and OTA cross the shell and are mostly found at the level of 'carbohydrate' substrate which almonds represent. The localization of OTA depends therefore of the cocoa beans state (integrity of the shell, bean of good quality).

OCCURRENCE OF OTA IN DERIVED COCOA PRODUCTS

Titration of OTA in Cocoa Butter

Samples of cocoa butter extracted are less contaminated by OTA: 25 % of butter samples have an OTA content superior to the limit of detection

(0.03 ng.g^{-1}) and 67 % of butter samples have an OTA content superior or equal to European tolerable limit, defined for derived cocoa products which is 1 ng.g^{-1}. The maximum content titrated was 1.88 ± 0.32 ng.g^{-1}. This content is superior to the norm. Butter samples contaminated by OTA are exclusively extracted from nibs obtained in damages cocoa pods and deferred pod-opening (T4). OTA is soluble in polar organic solvents, in diluted solution of sodium bicarbonate, moderately soluble in water [41]. Therefore, during the cooking of cocoa paste, butter is liberated from oleiferous cells and floats on the surface. The biggest part of OTA contained in the sample will therefore remain in the solid fraction; only a very little part could be found in the lipid part.

Titration of OTA in Cocoa Powder

Cocoa powder produce have reduce lipid contents. A total of 36 samples of cocoa powder were produced and analyzed at HPLC. Sixty seven percent (67 %) of these samples are contaminated by OTA (content superior or equal to 0.03 ng.g^{-1}) and 50 % of positive samples have an OTA content superior to the tolerable set defined for cocoa derived products (1 ng.g^{-1}). The maximum content is 10.15 ng.g^{-1} during the 2005 cocoa season. With the cooking, a light drop of OTA content was observed: from 18.19 ng.g^{-1} in nibs to 16.15 ng.g^{-1} in cocoa powder providing from treatment T4 of the 2005 cocoa season. Generally, an average reduction of 10 % was observed in positive samples because of cooking. Cocoa powder constitutes one of the most contaminated fractions by OTA [7, 13, 18, 24]. Vecchio & Finoli [44] showed that 82% of cocoa powder commercialized in Italy was contaminated by OTA. During cooking, OTA remains in the solid phases. During this thermal treatment, it is possible that a complex is made by OTA and other carbon molecules present in the milieu [43]: this explains the reduction in OTA content of about 10 % observed.

Titration of OTA in Chocolate

Like in the cocoa powder, chocolate is one of the most contaminated cocoa derived products by OTA. Sixty seven (67 %) of chocolate samples are also contaminated by OTA, and 50 % have contents superior to the norm defined by the European Union. The maximum content titrated is 16.15 ng.g^{-1} observed on treatment T4 of the 2005 cocoa season.

A large preponderance of OTA in chocolate samples is revealed in the literature. It is the case of Burdaspal and Legarda in 2003 [11] who evaluated the occurrence of OTA in 296 samples of different types of chocolate in Spain and 15 other countries. OTA was detected in all the samples, except one (99.7 % of positive samples). Similar study was done by many other authors in Italy, German and other countries with the same results (large OTA content of chocolate sample) [24, 32].

INFUENCE OF THE PROCEDURES OF TRANSFORMATION ON THE LEVEL OF OTA IN PRODUCTS

Thirty six (36) samples of cocoa beans coming from several types of post-harvest treatments were transformed by traditional techniques into cocoa butter, cocoa powder and chocolate paste. This was done in order to determine the influence of the procedure of transformation on OTA content of process products. Among 36 samples of cocoa beans analyzed, 12 samples had OTA contents superior to the threshold tolerable for cocoa beans which is 2 ng.g^{-1}. Only these beans are taking into consideration to study the impact of procedures on OTA content. Table 4 presents the mean contents of contaminated beans and derived products.

Table 4: Mean content of OTA in contaminated nibs and derived products

Samples	OTA content (ng.g^{-1})	Mean content (ng.g^{-1})
Cocoa beans	5.34	11.52
	20.75	
	10.0	
	10.0	

Roasted Cocoa	5.12 19.86 10.11 10.11	11.30
Nibs	4.08 18.19 9.45 8.22	9.98
Cocoa Butter	0.02 1.88 1.10 0.93	0.98
Cocoa powder	1.55 16.15 7.22 6.99	7.97
Chocolate spread	1.95 17.00 8.89 7.33	8.79

Figure 1 gives the influence of traditional procedures on OTA content in transformed products. This figure shows that derived cocoa products conserve a high contamination by OTA. Neither shelling nor traditional procedures have contributed to reduce the OTA content of products. Only cocoa butter is lightly contaminated (0.98 $ng.g^{-1}$). Cocoa powder and chocolate spread have a proximally the same OTA mean content (about 8 $ng.g^{-1}$). Similar works were done by Dembele and al. in 2009 [14] in Ivory Coast. This study was made on 16 samples of cocoa with more important levels of OTA (contents ranged between 3.37 and 46.15µg.kg-1). These samples were transformed using modern techniques into derived cocoa products (cocoa butter, chocolate...). In average, about 70% of OTA was taken out with the shell fraction and processed products obtained were lightly contaminated by OTA. Chocolates contained 1.86 µg.kg-1in averages and butter was free from

OTA. Because of the fragility of bean pellicle due to injured pods and deferred pod-opening, contamination by OTA is not longer superficial and 'shelling does not longer contribute to reduce significantly OTA content in beans. This therefore explains important OTA contents of derived cocoa products. Traditional procedures of cocoa processing do not have an influence on OTA content in derived products.

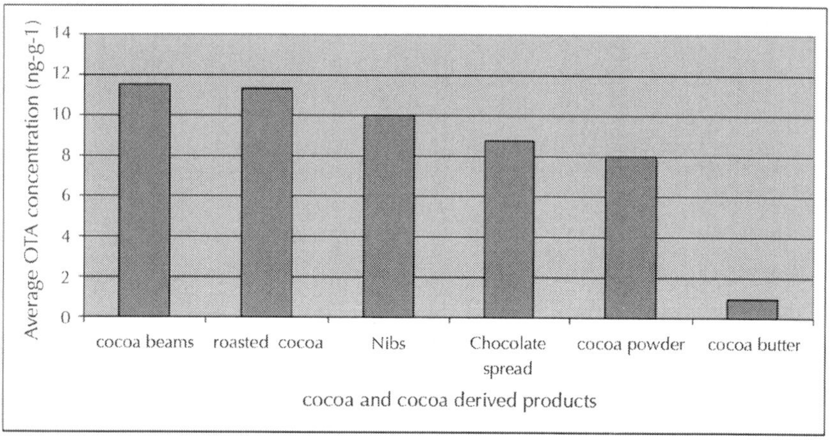

Figure 1: Influence of traditional procedures on Ochratoxin A (OTA) content in transformed products...

REFERENCES

1. L. Abrunhosa, R.R.M. Paterson,, Kozakiewicz., N. Lima, A. Venancio, (). Letters in applied microbiology, 2001,32: 240-242.
2. Adewole E, Ogunmodede O.T1, Talabi J 1, Ajayi O.O. 1 ,Oso O.A 1 and Lajide L. J. Chem. Pharm. Res., 2011, 3(6):544-552
3. S. Amézqueta, E. Gonzalez-Penas, M.Murillo, A. Lopez de Cerain, Food Add. and Conta. 2005, 22: 590-595.
4. E.K Asep, S. linap, T.J. Tan, A.R. Russly, S. Harcharan, and S.A.H. Nazimah, J. of Food Engineering, 2008, 85: 450-458.
5. M. Bakker, M. N. Pieters. Risk assessment of ochratoxin A in the Netherlands. RIVM report 388802025, 2002

6. P. Bastide, G. Fourny, N. Durand, P. Petithuguenin, B. Guyot, M. Gilmour, M. Lindblom. 15th Intl Cocoa Res. Conf., San Jose, Costa Rica, 2006, 9-17 October.
7. S.J. Bonvehi. J. Agric. Food chem., 2004, 52: 6347-6352.
8. H. Boudra, P. Le Bars, , J. Le Bars. Appl. Environ. Microbiol., 1995, 61:1156-1159.
9. C. Brera, R. Caputi, , M. Miraglia, , I. Iavicoli, , A. Salerno, , & G. Carelli. Microchemical Journal, 2002, 73 :167-173.
10. C. Brera, F. Debegnach, B. De Santis, E. Iafrate, E. Pannunzi, C. Berdini, E. Prantera, E. Gregori, M. Miraglia. Food Control 2011, 22: 1663-1667
11. P.A. Burdaspal, and T.M. Legarda,. Alimentaria, 2003, 40: 143-153.
12. E. Chiavaro, A. Lepiani, F. Colla, P.Bettoni, E.Pari, E. Spotti, , (2002). Food Add. and Conta., 19: 575-581.
13. M.V. Copetti. Micobiota do cacao: Fungos e Micotoxinas do cacau ao chocolate. PhD. Thesis, Faculdade de Engenharia de Alimentos da Universidade Estadual de Campinas, Campinas, SP. 2009
14. A. Dembele, A.Coulibaly, S.K. Traoré, K. Mamadou, ; Silue, N. ; Abba Touré, A. Tropicultura, 2009, 27: 26-30.
15. G. Engel. Archiv für lebensmittelhygiene, 2000, 51: 98-101.
16. M.B.M., Ferraz, A. Farah, , B.T. Iamanaka, D, Perrone, M.V, Copetti, , V.X. Marques, A.A. Vitali, and M.H. Taniwaki. Food Control. 2010, 21:872-877.
17. M. Gilmour, , & M. Lindblom. Management of ochratoxin A in the cocoa supply chain: a summary of work by the CAOBISCO/ECA/FCC working group on ochratoxin A. In J. F. Leslie, R. Bandyopadhyay, & A. Visconti (Eds.), Mycotoxins: Detection methods, Management Public Health and agricultural Trade 2008, (pp. 231-243). Wallingford, UK: CABI.
18. M. Gilmour, and M. Lindblom. MYCOGLOBE Conference:. 2005, Accra, Ghana, 13 – 16 http://www.iita.org/mycotoxinconf/index.htm
19. D. Höhler. Zeitschrift für Ernährungswissenschaft, 1998, 37: 2-12.

20. J. K. Koka*, 1D. K. Dodoo, 2P. K. Kwakye and 1J. Kambo- Dorsa. J. Chem. Pharm. Res., 2011, 3(4): 467-471

21. T. Kuiper-Goodman,. Food add. and conta. 1996, 13: 535-557.

22. T. Kuiper-Goodman, P.M. Scott. Biometrics biomedical and environment sciences, 1989, 2, 179-248.

23. W, Krysiak, Journal of Food Engineering. 2006, 77 : 449-453.

24. MAFF. Food Surveillance information Sheet No.130. 1999, http://www.archive.food.gov.uk/maff/food/infsheet/1999/no185/185ochra.htm

25. P.G. Mantle., K.M. McHugh. Mycological research, 1993, 97: 205-212.

26. P. Mounjouenpou, N. Durand, B. Guyot, P.J. Guiraud. Food add. and conta. 2007,7: 730-734

27. P. Mounjouenpou, D. Gueule, B. Guyot,. P.R. Tondje,. A. Fontana-Tachon, J.P.Guiraud. Int. J of Food Microbiol 2008, 121 : 234 – 241.

28. P. Mounjouenpou, D.Gueule, M. Ntoupka, N. Durand, A. Fontana-Tachon, B. Guyot, J.P Guiraud, World Mycotoxin Journal: 2011, 2: 141-146.

29. C. Micco, A. Ambruzzi, M. Miraglia, C. Brera, R.,Onori, L Benelli. IARC Science Publication, 1991 Vol. 115 (pp. 105-108)..

30. C. Micco, M. Miraglia, C. Brera, ,S. Corneli, A.Ambruzzi. Food Add.and Conta., 1995, 3: 351-354.

31. B. W Minifie, Science and Technology, 2nd ed. AVI Publishing Company, Westport, Connecticut. 1982

32. M. Miraglia, C. Brera. Reports on tasks for scientific cooperation, task 3.2.7, 69-86. Publisher: SCOOP, Directorate-General Health and Consumer Protection 2002

33. http://www.europa.ev.int/comm/food/fs/scoop/3.2.7_en.pdf

34. M. Miraglia, A. de Dominicis, , C. Brera, S. Corneli, E. Cava, E. Menghetti, M., Miraglia.. Natural Toxins, 1995, 3, 436-444.

35. M. Nakajima, H. Tsubouchi, M. Miyabe, , Y. Ueno. Food and Agri. Immunology, 1997, 9 : 77–83.

36. T. N. T. Nwagu,. F.S. Ire. Int. Jour. of Agri. Res., 2011, 6:101-118.

37. O'Callaghan J., Caddick M.X., Dobson D.W. Microbiology, 2003, 149, 3485-3491.

38. E. Owusu-Ansah1, 2*, J.E. Koranteng-Addo2, L.K. Boamponsem1, E. Menlah1, E.

39. Abole. J. Chem. Pharm. Res., 2010, 2(4):580-587

40. D. Palli, M. Miraglia, C. Saieva, G. Masala, E. Cava, M. Colatosti.. Cancer Epidemiology, Biomarkers & Prevention, 1999, 8: 265-269.

41. [39] K. Thirumala-Devi, M.A. Mayo, G. Reddy, K.E Emmanuel,. Y. Larondelle, D.V.R. Reddy. Food Add. And Conta., 2001, 18 : 830–835.

42. J.I Pitt,. J.C. Basilico, M.L. Abarca, C.Lopez, Medical Mycology, 2000, 38 : 41-46.

43. P.M. Scott,.Food Add. and Conta.. Fourth edition. 1996,Elsevier, Amsterdam. pp.214-220.

44. M. Suàrez-Quiroz,, O. Gonzàlez-Rios, M. Barel,. B. Guyot, S. Schorr-Galindo, J.P. Guiraud, Journal of Food Science and Technology, 2004, 39 : 501–507.

45. M. Suàrez-Quiroz., O. Gonzàlez-Rios, M. Barel, B. Guyot, , S. Schorr-Galindo, J.P. Guiraud.. Int. Journal of Food Microbiol, 2005, 40 : 605-611.

46. A. Vecchio,. C. Finoli. Industrie Alimentaria, 2007, 46:1015-1020.

Chapter 7

Factors Influencing Adoption of Agricultural Processing Technologies Developed by National Centre for Agricultural Mechanization (NCAM) in Ifelodun Local Government Area, Ilorin Kwara State

Mohammed, B. T., Achem, B. A., and Abdulquadri, A. F.

National Centre for Agricultural Mechanization (NCAM) Km 20, Ilorin-Lokoja Highway, Idofian, P.M.B. 1525, Ilorin, Kwara State, Nigeria

ABSTRACT

This study was carried out in Ifelodun Local Government Area of Kwara State, Nigeria with the objective to promote adoption of NCAM agricultural processing technologies among farmers in the neighboring communities. Data was collected using a 2-stage sampling technique randomly selected from 77 respondents. Interview schedule with a structural questionnaire was used to collect data for the study. The data were analyzed using simple descriptive statistics and multiple regression analysis. The results from the findings indicated that adoption of NCAM developed processing technologies is majorly in the hands of female, middle aged, educated and full-time farmers. Also, majority (about 67.5%) did not belong to any farmers group or cooperative society, 70% were not aware of NCAM technologies. The multiple regression analysis result indicates that the co-efficient of sex, cooperative society, relevant aspiration as well as age were fund to be significantly influencing the adoption of these two NCAM technologies. The study recommends good finishing, durability of materials, need for adequate publicity, feedback, and provision of adequate extensionist in the study area.

INTRODUCTION

In Nigeria, agriculture plays a vital role in economic development. This is because its contributes about 37.2% of the country's Gross Domestic Product (GDP), provides direct employment to about 75% of the population (NBS, 2007, Falusi, 2008), accounts for only 44% of Nigeria's total exports and 56% of total non-oil export, provides over 90% of the food requirement of the Nigerian population.

Agriculture also provides raw materials for 60% of industries in the country. In spite of all these tremendous potential, it is indeed not only an irony but an embarrassment, that Nigeria, a country endowed with substantial natural resources and ecological diversities still find itself in the unenviable group of low income and food deficient countries in Africa. Any attempt at addressing the issues of food security in Nigeria must of necessity concentrate on efforts to stimulate adoption of technologies among our peasant farmers who remains the main driver of agricultural production.

According to Abalu (1993) "why they may not be general consensus, it can nonetheless, be concluded that technology stocks capable of attaining growth rates in excess of 8.0 – 10.0 percent do in fact exist or can be made available for several agricultural commodities in the various ecological Zones". This statement is particularly applicable to the National Centre for Agricultural Mechanization (NCAM) which has made tremendous efforts in the design and development of simple agricultural technologies aimed at improving the income and livelihood of small scale farmers.

NCAM has made some modest breakthrough in the development of a peeling tool for cassava (Manihot esculaenta) a crop considered to be very important in Nigeria. Cassava is important as a major source of food, providing close to 70% of the calorific requirement of the Nigeria diet (Okigbo, 1980). Cassava is also important in the Industrial Sector where it is used for confectionary baking, animal feeds, pharmaceuticals, ethanol, textile, biofuel etc. Unfortunately, mechanization of cassava peeling has proved to be a major challenge in cassava processing due to lack of size uniformity and configuration.

Okro (Abelmoschus esculentus), is a vegetable plant widely consumed in Nigeria and has a potential to improve nutrition, boost food security, foster rural development and sustainable land care (NRC. 2006). Okro is consumed fresh where the mucilaginous slime contains usable form of soluble fibre popularly called Draw Soup in Nigeria. In some cases, the fresh fruit is sliced, dried and pounded to produce soup. Okro leaves may also be cooked in a similar way to the green vegetable or eaten raw in salad (Miiliiken and Feniger, 1996). A characteristic of fruits and vegetables is their tendency to spoil and get wasted if not processed within a short period. In Nigeria and Kwara State in particular, large quantities of this important vegetables are produced during the rainy season and most of them are not processed and preserved. Consequently, the prices of the vegetable seem to be uncompetitive low during the peak period and become very scarce and expensive during the dry season. NCAM has similarly developed a simple technology that can be used for slicing okro for drying and preservation in order to improve the income and food security of farmers.

The adoption rate of simple processing technologies developed by NCAM like cassava peeling tool and okro slicer remain very low thus limit the purpose of which they are meant for. Adoption of agricultural technologies remains a viable strategy for making farmers economically viable and increase farm productivity. In spite the effort of NCAM in development of simple processing technologies, there is dearth of information on the adoption of these technologies.

The aim of this study is to bridge that gap. Specifically, the study was designed to:

- Describe the socio-economic characteristic of respondents in the study area,
- Determine the factors affecting adoption of agricultural developed technologies in the study area.

METHODOLOGY

This study was carried out in Ifelodun Local Government Area (LGA) of Kwara State, Nigeria. Ifelodun is the largest Local Government Area in Kwara State with an estimated population of about 206,042 and an estimated total land area of about 3,435Km^2 (NPC, 2006 and KWSMI, 2002). The area is located between latitude 70 45′N and 90 30′ N and longitude 20 30′E and 60 35′E.

It is characterized by dry and wet season. The annual rainfall ranges between 1000mm and 1500mm. Average temperatures varies between 300 c and 350 c while humidity range from 35% to 60%. The major source of livelihood and occupation of the people in the area is farming. Farming is traditional in nature with emphasis on the cultivation of crops such as cassava, yam, maize, sorghum and vegetables like Okro, Amaranthus etc (KWSMI, 2002 and Mohammed, 2008). Cassava and Okro are predominantly grown in the area; this informed the choice of the National Centre for Agricultural Mechanization (NCAM) to develop processing prototype machines for these crops in thestudy area.

A 2-stage sampling technique was employed for the study. The first stage involved the purposive selection of four communities namely; Ganmo, Fufu, Idofian and Jimba-Oja. Based on their relative proximity to the Centre. In the second stage, a random sampling technique was used to select 77 farmers from the selected communities based on their relative population size. Preliminary meetings were held with the village/community opinion leaders to sensitize them on their relevance in mobilizing the farmers in their domain for the adoption of NCAM develop processing technologies.

The technologies to be adopted were demonstrated to the farmers while arrangement was made for publicity, press coverage (Radio jingles) during the demonstration after which the farmers were provided with the samples for their private use. The data used for this study was basically primary data. This involved the use of an interview schedule with a structured questionnaire administered to the farmers.

Relevant data collected include; socio-economic characteristics, sources of information on research results/technologies, adoption of technologies and constraints militating against adoption of NCAM processing technologies.

The data collected from this study was analyzed using descriptive and quantitative methods. This includes the use of percentages, frequencies and averages. Multiple regression analysis was used to determine factors affecting adoption of agricultural developed product in the study area.

The relationship between the adoption of agricultural technologies and factors affecting it can be expressed implicitly as follows:

$$Y = f(X_1, X_2, X_3, X_4, X_5, X_6, X_7, X_8, X_e)$$

where:

Y = adoption of NCAM technologies

X_1 = age of farmers

X_2 = Sex

X_3 = Marital status

X_4 = Experience

X_5 = Family size

X_6 = Source of Information

X_7 = Educational status of farmers

X_8 = Association among farmers

E = Error Term

Chapter 8

Effect of Moisture, Bulk Density and Temperature on Thermal Conductivity of Ground Cocoa Beans and Ground Sheanut Kernels

A. Bart-Plange, A. Addo, S. K. Amponsah & J. Ampah

Kwame Nkrumah University of Science and Technology, Kumasi, Ghana

ABSTRACT

Thermal conductivity is an important engineering parameter in the design of food processing equipment. It predicts or controls the heat flux in food during processing such as cooking, frying, freezing, sterilization, drying or pasteurization. The thermal conductivity of ground cocoa beans and ground sheanut kernels with varying moisture content, bulk density and temperature was studied using the transient heat transfer method. The thermal conductivity increased linearly for ground cocoa beans sample from 0.0243 to 0.0311 W/oCm and for ground sheanut kernels from 0.0165 to 0.0458 W/oCm in the moisture content range of 12.59 to 43.84 % w.b. at a constant bulk density of 295 kg/m3. For bulk density range of 322 to 410 kg/m3, thermal conductivity of ground cocoa beans and ground sheanut kernel increased linearly from 0.0265 to 0.0324 W/oCm and 0.0209 to 0.0252 W/oCm respectively when moisture content was at 16 % w.b Thermal conductivity of ground sheanut kernel and ground cocoa beans increased significantly ($p<0.05$) from 0.0233 to 0.0382 W/oCm and 0.0261 to 0.0397 W/oCm respectively as temperature increased from 35 to 55 °C. Effect of moisture, bulk density and temperature on thermal conductivity of sheanut kernel and cocoa bean were found to be significant ($p>0.05$).

INTRODUCTION

Sheanut and Cocoa are important oil producing crops in Ghana. Sheanut hails from the Sapotaceae family. The commonly known varieties include Vitellariaparadoxa (Butryospermum parkii) and Vitellarianilotica. Cocoa (Theobroma Cacao) is an ancient crop of the lowland tropical forest, which originated from the Shea nut is obtained from the shea tree, and is grown mostly throughout West and Central Africa; in the semiarid Sahel from Senegal to Ethiopia (Aremu and Nwannewuihe, 2011). Shea nut contains reasonably high amounts of oleic acids from which the shea butter (fat) is obtained. Shea butter is one of the basic raw materials for most food, cosmetics, soap as well as the pharmaceutical industries (Thioune et al., 2000) and it is sometimes used as a substitute for cocoa butter (Bekure et al., 1997).

Southern and Central America (Lefeber et al., 2011). In West Africa, cocoa is one of the most important cash crops. Globally, Ghana's cocoa bean production is ranked second in the world after her western neighbour Côte d'Ivoire (FAOSTAT, 2005). Ghana is recognized as the world leader in premium quality cocoa beans production. Cocoa serves as the major source of revenue for the provision of socio-economic infrastructure in the country. In terms of employment, the industry employs about 60% of the national agricultural labour force in the country (Appiah, 2004 cited in Ntiamoah and Afrane, 2008).

For these farmers, cocoa contributes about 70 -100% of their annual household incomes (COCOBOD, 2004 cited in Ntiamoah and Afrane, 2008). Cocoa seeds are the source of commercial cocoa beans and cocoa products include cocoa liquor, cocoa butter, cocoa cake and cocoa powder as well as chocolate. Cocoa powder is essentially used as flavour in biscuits, ice cream and cakes and is consumed by most beverage industries.

Besides the traditional uses in chocolate manufacture and confectionery, cocoa butter, like shea butter, is also used in the manufacture of cosmetics. It is also a folk remedy for burns, cough, dry lips, fever, malaria, rheumatism and wounds. Studies show that the cocoa bean contains flavonoids with antioxidant properties that can reduce blood clot and the risk of stroke and cardiovascular attacks (ICCO, 2011).

The thermal conductivity of materials can be influenced by a number of factors such as the moisture content of the material, porosity and fibre orientation of the material (Stroshine and Hamann, 1994; Mohsenin, 1990). Thermal conductivity of food and biological materials increase with increase in moisture content and density (Opoku et al., 2006 for hay; Muramatsu et al. 2006 for brown rice, Aviara et al., 2008 for guna seeds and Perusella et al., 2010 for banana). Thermal conductivity data is needed for calculating energy demand for the design of equipment and optimization of thermal processing of foods (Polley et al., 1980). It controls the heat flux in food during processing such as cooking, frying, freezing, sterilization, drying or pasteurization. In the determination of thermal conductivity of food materials, the commonly employed methods arethe transient and the steady-state methods (Mohsenin, 1980). Besides processing and preservation, thermal conductivity and other properties such as specific heat and thermal diffusivity also affect sensory quality of foods as well as energy saving during processing (Opoku et al. 2006).It is not uncommon to see farmers dry their produce without taking into consideration the quantity of heat needed to accomplish the drying process which in turn affects the market value of the end product. This is because such information on thermal conductivity of local agricultural products is either unavailable or inadequate.

The objective of this study therefore, was to determine the thermal conductivity of ground sheanut kernel and ground cocoa beans and investigate their dependence on moisture content, bulk density and temperature using the transient heat method.

MATERIALS AND METHODS

Sample Preparation

The samples were cleaned by removing foreign materials and damaged kernels or beans. Sheanut kernel and cocoa beans had all the quality checks performed and ready for local and export market. Both samples were conditioned to four moisture content levels of 12.59, 22.41, 31.55 and 43.84% wet basis. The samples were sealed in separate polythene bags and kept in a refrigerator at 5°C for five days to ensure uniform moisture distribution. The amount of distilled water added was calculated using equation (1) (Balasubramanian, 2001).

$$M_w = \frac{M_i(m_f - m_i)}{100 - m_f} \tag{1}$$

where:

M_w is the mass of distilled water (g),

M_i is the initial mass of sample (g),

m_f is the final moisture content of sample (%w.b.) and

m_i is the initial moisture content of sample (%w.b.).

For bulk density variation sampleswere milled in a laboratory hammer mill to a particle size of two millimetres using a set of screen. Varying bulk densities of 322, 346, 381 and 410 kg/m3 were obtained by compressing in a cylinder with known weights at a constant moisture content using standard procedures (AOAC, 2002).

Experimental Setup

The setup for the thermal conductivity measurements is shown in Figure 1. The thermal conductivity apparatus is a set-up consisting of an aluminium cylinder with a heating coil stretching between two insulated ends of the cylinder. A thermocouple was fitted through the top end of the cylinder for temperature readings in the sample. Heat source (Q) was supplied by a constant direct current power source with current and voltage of 1A and 3V respectively throughout the experiment. In the set-up, there was an ammeter to take current readings, voltmeter to take voltage readings and a rheostat to vary resistance in the circuit in order to achieve the desired current.

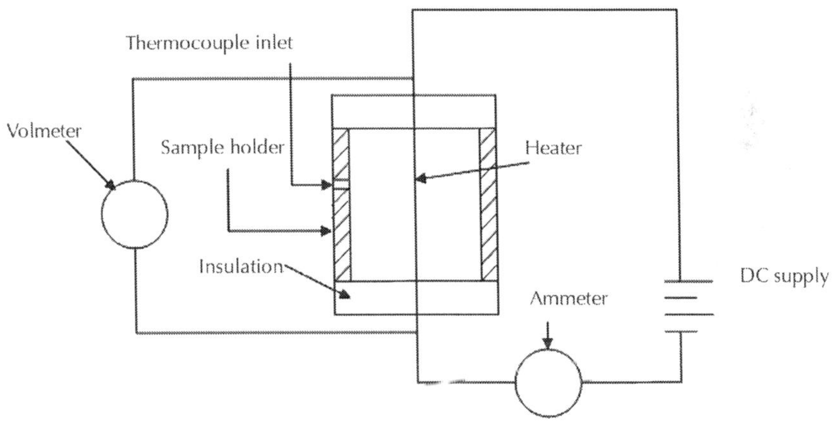

Figure 1: Schematic diagram of the thermal conductivity measuring apparatus

The conditioned samples were placed in the aluminium sample cylinder in the set-up. The sample temperature at the centre of the cylinder was checked by means of a thermocouple, the current was adjusted to one ampere and a voltage of three volts was used. Temperature readings were taken at regular time interval of one minute for 40minutes for each sample experimented. Thermal conductivity was determined at five sample temperatures of 35, 40, 45, 50 and 55°C.

Thermal Conductivity Determination

Thermal conductivity (k) was determined using equation (2) (Tabil, 1999).

$$k = \frac{Q}{4\pi(t_2 - t_1)} \ln \frac{\theta_2}{\theta_1} \quad (2)$$

where:

Q = VI (V is the supplied voltage, I is input current, heat source per meter of the line source)

k = thermal conductivity of the medium (W/°Cm)

t_1 is initial temperature (°C)

t_2 is final temperature (°C)

θ_1 is initial time (min)

θ_2 is initial time (min)

The thermal conductivity values of the ground sheanut kernel and ground cocoa beans were determined by calculating the slopes of the graphs of temperature changes against time ratio on a semilogarithm graph. The experiment was replicated four times at each moisture content, bulk density and temperature level and thermal conductivity was recorded in each case.

Statistical Analysis

The experimental design used was the completely randomized design (CRD) with single factor analysis of variance (ANOVA) for all data and analyzed with Minitab Version 15.Statistical significance was carried out using Tukey and Fisher's approach at $p<0.05$.

RESULTS AND DISCUSSION

Effect of Moisture Content

Figure 2 shows the linear variation of thermal conductivity of ground sheanut kernel and ground cocoa beans with moisture content at constant bulk density (295 kg/m^3). Thermal conductivity of ground sheanut kernel and ground cocoa beans increased significantly ($p<0.05$) from 0.0165 to 0.0458 W/°Cm and 0.0243 to 0.0311 W/°Cm respectively with increasing moisture content from12.59 to 43.84 %w.b.

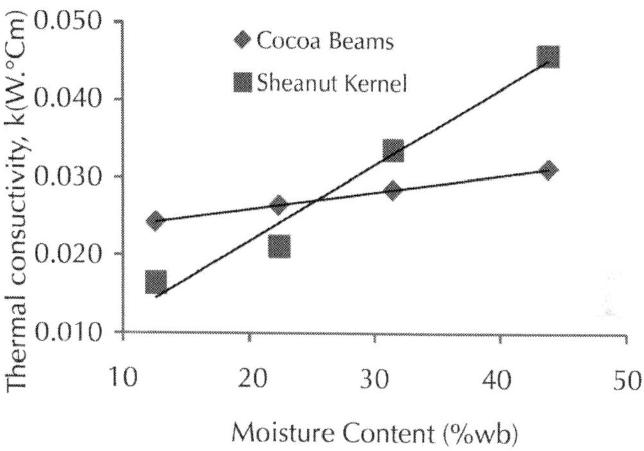

Figure 2: Thermal conductivity as a function of moisture content

Similar trend was observed in the thermal conductivity of soybean (Deshpande et al., 1996), cumin seed (Singh and Goswami, 2000), sheanut kernel (Aviara and Haque, 2001), borage seed (Yang et al., 2002), millet grains (Subramanian and Viswanathan, 2003), rough rice (Yang et al., 2003), brown rice (Muramatsu et al., 2006), maize and cowpea (Bart- Plange et al., 2009) and guna seed (Aviara et al., 2008). The increase in thermal conductivity with moisture content can be attributed to the fact that an increase in moisture content of the sample increases the amount of water molecules available to fill the pores within the sample thus increasing the ability of the sample to conduct more heat.

The relationship between the thermal conductivity of ground sheanut kernel (k_{sk}) and ground cocoa beans (k_{cb}) and moisture content (M) can be expressed using equations (3) and (4) respectively.

$$k_{sk} = 0.097M + 0.002, R^2 = 0.974 \quad (3)$$

$$k_{cb} = 0.022M + 0.021, R^2 = 1 \quad (4)$$

Equations (3) and (4) depict that thermal conductivity of ground sheanut kernel and ground cocoa beans have a linear relationship with moisture content. This agrees with what was reported by Aviara and Haque, (2001) on sheanut kernel. Studies by other researchers also found thermal conductivity to increase with increasing moisture content (Sweat, 1974; Rao and Rizvi, 1986; Mohsenin, 1990; Tansakul and Lumyong, 2008; Meghwal and Goswami, 2011).

Effect of Bulk Density

The variation of thermal conductivity of ground sheanut kernel and ground cocoa beans with bulk density at constant moisture content (16 %w.b.) is shown in Figure 3.

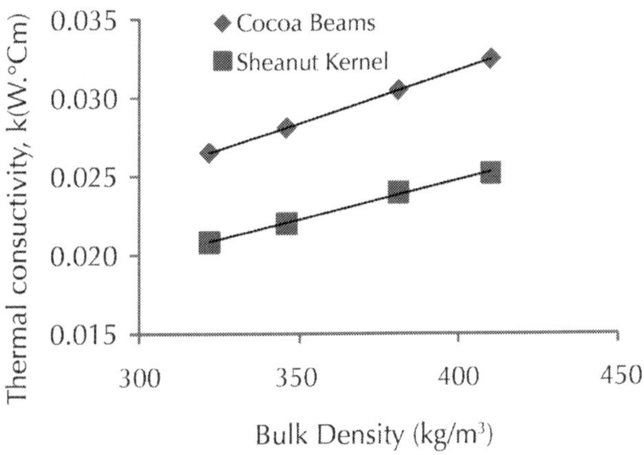

Figure 3: Thermal conductivity as a function of bulk density

The thermal conductivity of ground sheanut kernel and cocoa beans increased linearly and significantly ($p<0.05$) from 0.0209 to 0.0252 W/°Cm and 0.0265 to 0.0324 W/°Cm respectively as bulk density increased from 322 to 410 kg/m^3. This trend was similarly observed by other researchers including Taiwo et al (1996) for cowpea, Aviara and Haque (2001) for sheanut kernel, Bart-Plange et al. (2009) for maize and cowpea and Meghwal and Goswami (2011) for black pepper. Moreover, it was observed that the thermal conductivity of ground sheanut kernel was generally lower than ground cocoa beans with increase in bulk density. The increase in thermal conductivity with bulk density can best be explained by making reference to the conduction ability of the sample particles in relation to the pores between them.

Increasing the bulk density means increasing the number of particles in a constant volume thus decreasing the pore volume which leads to increased heat conduction ability of the sample.

The linear relationship between the thermal conductivity of ground sheanut kernel (k_{sk}) and ground cocoa beans (k_{cb}) and bulk density (ρ) may be expressed using equations (5) and (6) respectively.

$$k_{sk} = 5\times10^{-5}\rho + 0.004,\ R^2 = 0.997 \quad (5)$$

$$k_{cb} = 7\times10^{-5}\rho + 0.004,\ R^2 = 1 \quad (6)$$

Effect of Temperature

Figure 4 describes the variation of thermal conductivity with increasing temperature at constant moisture content and bulk density. At constant moisture content of 20.5 %w.b. and constant bulk density of 420 kg/m³, the thermal conductivity of ground sheanut kernel and ground cocoa beans increased significantly ($p<0.05$) from 0.0233 to 0.0382 W/oCm and 0.0261 to 0.0397 W/oCm respectively as temperature increased from 35 to 55 °C.

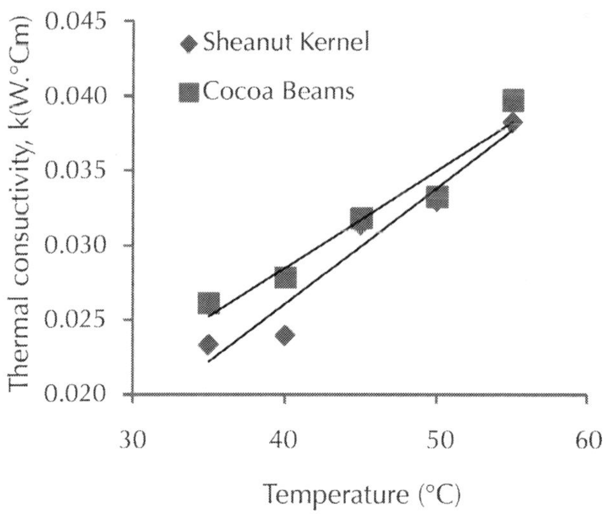

Figure 4: Thermal conductivity as a function of temperature

The linear relationship between the thermal conductivity of ground sheanut kernel (k_{sk}) and ground cocoa beans (k_{cb}) and temperature (°C) can be expressed using equations (7) and (8) respectively.

$$K_{sk} = = 0.0008x - 0.005 \quad R^2 = 0.9446 \tag{7}$$

$$K_{cb} = 0.0007x + 0.0024 \quad R^2 = 0.9434 \tag{8}$$

Aviara and Haque (2001) and Bart-Plange et al. (2009) observed an increase in thermal conductivity with temperature for sheanut kernel and maize and cowpea respectively. Kurozawa et al. (2008) found thermal conductivity to increase from 0.57 to 0.61 W/m °C with temperature in the range of 25 to 45 ºC for cashew apple. Mahmoodi and Hosein (2008) also found thermal conductivity of pomegranates to increase linearly from 0.6106 to 0.6372 W/m°C with increase in temperature from 26.5 to 45°C.

CONCLUSION

Investigations on the thermal conductivity of ground sheanut kernel and ground cocoa beans revealed the following:

1. Thermal conductivity of ground sheanut kernel and ground cocoa beans increased significantly ($p<0.05$) from 0.0165 to 0.0458 W/oCm and 0.0243 to 0.0311 W/oCm respectively with increasing moisture content from 12.59 to 43.84 % w.b. at constant bulk density. A linear relationship was found to exist between thermal conductivity and moisture content.
2. Thermal conductivity of ground sheanut kernel and cocoa beans increased significantly ($p<0.05$) from 0.0209 to 0.0252 W/oCm and 0.0265 to 0.0324 W/°Cm respectively as bulk density increased from 322 to 410 kg/m^3 at constant moisture content. A linear regression best describes the relationship between thermal conductivity and bulk density.
3. At constant moisture content of 20.5 %w.b. and constant bulk density of 420 kg/m^3, the thermal conductivity of ground sheanut kernel and ground cocoa beans increased significantly ($p<0.05$) from 0.0233 to 0.0382 W/oCm and 0.0261 to 0.0397 W/oCm respectively as temperature increased from 35 - 55 °C.

REFERENCES

1. AOAC. (2002). Official Methods of Analysis, 17th Ed. Association of Official Analytical Chemists, Gaithersburg, Maryland, USA.
2. Appiah, M. R. (2004). Impact of cocoa research innovations on poverty alleviation in Ghana. Ghana Academy of Arts and Sciences Publication cited in Ntiamoah, A. and Afrane, G. (2008). Environmental impacts of cocoa production and processing in Ghana: life cycle assessment approach, Journal of Cleaner Production. 16(16): 1735-1740.
3. Aremu, A. K. and Nwannewuihe, H. U. (2011).Specific heat of ground fresh sheanut kernel (Butyrospernum paradoxum) as affected by particle size, moisture content and temperature. Journal of Emerging Trends in Engineering and Applied Sciences, (JETEAS) 2 (1): 177-183.
4. Aviara, N. A. and Haque, M. A., (2001). Moisture dependence of thermal properties of sheanut kernel. J. Food Eng., 47, 109-113.
5. Aviara, N. A., Haque, M. A. and Ogunjimi, L. A. O. (2008). Thermal properties of Guna Seed. Institute of Agrophysics, Polish Academy of Sciences Vol. 22: 291-297.
6. Balasubramanian, D. (2001). Physical Properties of Raw Cashew Nut. Journal of Agricultural Engineering Research. Vol. 78(3): 291 – 297.
7. Bart-Plange, A., Asare, V. and Addo, A. (2009). Thermal conductivity of maize and cowpea. Journal of Engineering and Technology. 2(3): 6-11.
8. Bekure, Z., Donlan, M., Gordon, Y. and Thomson, J. (1997). "Local to Global. The International market for Shea butter." Presented to United Nations Development Fund for Women.
9. COCOBOD (2004). Socio-economic study. Final report [MASDAR]. Ghana Cocoa Board cited in Ntiamoah, A. and Afrane, G. (2008). Environmental impacts of cocoa production and processing in Ghana: life cycle assessment approah, Journal of Cleaner Production. 16(16): 1735-1740.
10. Deshpande, S. D., Bal, S., and Ojha, T. P. (1996). Bulk thermal conductivity and diffusivity of soy bean. J. Food Process Preservation. 20: 177 189.

11. FAOSTAT (2005). Faostat database. Food and Agriculture Organisation.[Accessed 10 October, 2011]. Available online: www.fao.org/es/ess/top/ commodity. html?lang=en&item=125& year=2005.

12. ICCO, International Cocoa Organisation. (2011). Also available at www.icco.org. [Accessed September 1 2011].

13. Kurozawa, L. W., Park, K. J. and Azonbel, P. M. (2008). Thermal conductivity and thermal diffusivity of papaya (Carica papaya,L.) and Cashew apple (Anacardium occidentale L.). Braz. J Food Tech.11:78-85.

14. Lefeber, T., Gobert, W., Vrancken, G. Camu, N. and Vuyst, L. D. (2011). Dynamics and species diversity of communities of lactic acid bacteria and acetic acid bacteria during spontaneous cocoa bean fermentation in vessel. Food microbiology. 28: 457- 464.

15. Mahmood, M. and Hosein, K. M. (2008). Determination and Comparison of Thermal Conductivity of Iranian Pomegranate Varieties. 18th National Congress on food technology, Mashhyad, IR, Iran, 15-16 October , 2008.

16. Meghwal, M. and Goswami, T. K. (2011). Thermal properties of black pepper and its volatile oil. International Journal of Advanced Biotechnology and Research. 2(3): 334-344.

17. Mohsenin, N. N. (1980). Thermal Properties of Foods and Agricultural Materials. pp 83–121, Gordon and Breach, New York.

18. Mohsenin, N. N. (1990). Thermal properties of foods and agricultural materials. New York: Gordon and Breach.

19. Muramatsu, Y., Tagawa, A., Kasai, T., Takeya, K., and Fukushima, M. (2006). Prediction of thermal conductivity of kernels and packed bed of brown rice. Journal of the Japanese Society of Agricultural Machinery, 64(1): 70–76.

20. Opoku, A., Tabil, L. G., Crerar, B. and Shaw, M. D. 2006. Thermal conductivity and thermal diffusivity of timothy hay. Canadian Biosystem Engineering. 48: 3.1-3.7

21. Perusulla, C. A., Viviana, C., Mends, L. A. (2010). Development of a linear heat source probe and determination of banana thermal conductivity. International Journal of Food Engineering 6(5)

22. Polley, S. I., Synder, O. P. and Kotnour, F. (1980). A compilation

of thermal properties of foods. Food Technology. 34(11): 76–78.

23. Rao, M. A. and Rizvi, S. S. H. (1986). Engineering properties of foods. New York: Marcel Dekker.
24. Singh, K. K. and Goswami, T. K. (2000). Thermal properties of cumin seed. J. Food Eng. 45: 181- 187.
25. Stroshine, R., Hamann, D. D. (1994). Physical Properties of Agricultural Materials and Food Products. 1994. Course manual, Purdue University, USA.
26. Subramanian, S. and Viswanathan, R. (2003). Thermal properties of minor millet grains and flours. Biosys. Eng. 84: 289-296.
27. Sweat, V. E. (1974). Experimental values of thermal conductivity of selected fruits and vegetables. J. Food Sci. 39: 1080-1091.
28. Tabil, L. G., (1999). Specific heat of agricultural and food materials. Research report, Department of Agricultural and Bio-resource Engineering, University of Saskatchewan, Canada. p 7.
29. Taiwo, K. A., Akanbi, C. T., and Ajibola O. O. (1996). Thermal properties of ground and hydrated cowpea. J. Food Eng. 29: 249-256.
30. Tansakul, A. and Lumyong, R. (2008). Thermal Properties of Straw Mushroom. Journal of Food Engineering. 87:91-98.
31. Thioune, O., Ahodikpe, D., Dieng, M., Diop, A. B., Ngom, S. and Lo, I. (2000). Inflammatory ointment from shea butter and hydro alcoholic extract of Khaya sengalensis barks (cailcederat) Dakar Med. 45(2): 113-6.
32. Yang, W., Sokhansanj, S., Tang, J. and Winter, P. (2002). Determination of thermal conductivity, specific heat and thermal diffusivity of borage seeds. Biosys. Eng. 82: 169-176.
33. Yang, W., Siebenmorgen, T. J., Thielen, T. P. H and Cnossen, A. G. (2003). Effect of glass transition on thermal conductivity of rough rice. Biosystems Engineering. 84(2): 193-200.

Chapter 9

Effect Characteristics of Farmers on the Level of Technology Adoption Side-grafting in Cocoa Farming at Sigi Regency-Indonesia

Effendy[1,2], Nuhfil Hanani[3], Budi Setiawan[3] and A. Wahib Muhaimin[3]

[1]Agricultural Economics Faculty of Agriculture, University of Brawijaya, Indonesia

[2]Department of Agriculture Economic, University of Tadulako, Indonesia

[3]Department of Agriculture Economic, University of Brawijaya, Indonesia

ABSTRACT

This research aimed to determine the level of adoption and effect of characteristics of farmers on the level of adoption side-grafting technology on cocoa farming. The research was conducted at Palolo District, Sigi Regency-Indonesia. Sample size as big as 98 cocoa farmers and sampling was done by simple random. The results showed the level of technology adoption side-grafting on cocoa farming was still low. Characteristics of farmers such as ethnic, age, cocoa farming experience, and frequency followed agricultural extension of significant effect on the level of technology adoption side-grafting on cocoa farming at Sigi regency-Indonesia.

INTRODUCTION

Cocoa grew rapidly in the 1990s and making Indonesia the world's third largest exporter, after Ivory Coast and Ghana (Neilson, 2008). Cocoa farmers at Indonesia are now estimated to be 1,400,000 households, mostly small, about 2 hectares (ha) or less, even outside of Java. Cocoa prices increased very high during the economic crisis in the late 1990s actually have been brought blessing for cocoa farmers, especially at Eastern Indonesia. No exaggeration to say that the cocoa at Indonesia has contributed significant to poverty alleviation, especially in rural areas (Neilson, 2008). Cocoa is one of Indonesia commodity and plays an important role for the economy, particularly in the supply of employment and source of income. Indonesian cocoa market opportunities open enough, both domestic and export needs. Indonesian Cocoa Association (Askindo) estimated that in 2012 the domestic needs of cocoa beans increased by about 16.7 percent from last year (240,000 tons) to 280,000 tons. Cause of increase in the needs of cocoa beans, among others: corporate plan of Guan Chong Bhd increase the production capacity of its plant in Batam by 60,000 tons per year to 120,000 tons.

In addition, it was estimated, cocoa processing plant owned by PT Barry Callebaut Comextra Indonesia with of milling capacity 28,000 tons of cocoa beans per year in Makassar also has started operations in 2012 (Anonimous, 2012). Cocoa is one of the plantation commodities which have an important role in supporting economic development at Central Sulawesi-Indonesia that is as an export commodity. Cocoa production at Central Sulawesi in 2011 ranged from 95,589 tons (13.42% of Indonesian production) ranks third after South Sulawesi-Indonesia (20.07%) and Southeast Sulawesi-Indonesia (16.82%) (Direktorat Jenderal Perkebunan, 2012). Sigi Regency is one of cocoa producing areas at Central Sulawesi-Indonesia. But lately cocoa development is facing various problems such aslevel of crop productivity cocoa are still low. Causes of low productivity which is achieved cocoa of farmers is the presence of pests and diseases, the application of cultivation technology which is not optimal, and used plants type that have low production potential or conditions most plants have reached the age of 20 years (Basri, 2009).

To overcome these problems, the government through the Movement of Improvement Production and Quality National Cocoa (Gernas), has been implemented clonal propagation technology (type) superior cocoa, such as by side-grafting. Side-grafting in addition contribute to the multiplication of superior cocoa clones, also were effective in rehabilitating unproductive cocoa trees (old/broken) (Departemen Pertanian Direktorat Jenderal Perkebunan, 2009).

Side-grafting is a technique of propagation vegetative in which new plants that grow to be the same as the original plant. Stalk shoots/entres is taken from the most good tree of its growth and from selected clones, giving yields much, resistant to certain pests and diseases, as well as fit with the local environmental conditions, and has been applied by the Head of the Provincial Office in charge of plantations (Departemen Pertanian Direktorat Jenderal Perkebunan, 2009). When this has been known to some cocoa clones excels at Central Sulawesi-Indonesia, such as Sulawesi 1 (S1) and Sulawesi 2 (S2). Clones of cocoa S1 and S2 can be a source of budwood selection in order to increase the productivity of cocoa in Sigi regency-Indonesia through side grafting. Technology of side-grafting besides increase production also can stimulate fruitful cocoa. Normally cocoa produce fruit on 2-3 years, but for cocoa by technology of side-grafting has been able to bear fruit at the age of 1 year and production ranges from 1.8 to 2.75 tons$\cdot$ha^{-1} (Departemen Pertanian Direktorat Jenderal Perkebunan, 2009).

Side-grafting one of technology which is developed by farmers at Sigi Regency-Indonesia. Until the Year 2011 cocoa crop productivity was still around 0.67 tons·ha^{-1} with a total area of 27,555.10 ha and production of 18,386.50 tons (Badan Pusat Statistik, 2012). This is because not all farmers adopt the technology of side-grafting. The emergence technology of side-grafting as part of a sustainable farming system which is one of the answers to the reduction in productivity of cocoa in Sigi regency-Indonesia is very important to get serious attention. Technology of side-grafting in Sigi regency-Indonesia has not received full support from the government, researchers and farmers, so that is needed of steps strategic to communicate the technology to the farmers. The role of method approach in conveying an innovative so that farmers are willing to adopt the technology becomes very important to socialize side-grafting technology. Based on this context, the adoption of technology becomes very important to get in-depth attention. In relation with this thing, then it becomes interesting to do research on the effect of the characteristics of farmers to technology adoption level side-grafting on cocoa farming at Sigi regency-Indonesia. The purpose of the research is (1) determine the level of technology adoption side-grafting on cocoa farming and (2) analyze the effect of the characteristics of farmers to technology adoption side-grafting on cocoa farming.

METHOD

Place and Time of Research

The research was conducted at the Sejahtera Village and Bulili Village at Palolo Subdistrict-Indonesia. Both of village is the center of the largest cocoa production at Palolo Subdistrict and have been using side-grafting technology in plants of cocoa and has been produce. The study was conducted from January to March 2013. 2.2 Population and Sample The population in this research numbered 218 cocoa farmers. Population at Sejahtera Village 106 of cocoa farmers and at Bulili Village 112 of cocoa farmers. Samples was determined by probability sampling. Sampling was done by simple random. The number of samples was calculated by using the formula Parel et al. (1973), as follows:

$$n = \frac{N \sum N_h s_h^2}{N^2 \frac{d^2}{z^2} + \sum N_h s_h^2} \quad (1)$$

Where:

n = number of Sample N = number of population

N_h = number of population per village

d = precision is set at 10%

z = 1.645 (90%)

s_h = variance of each village

Number of samples from each village was determined proportionally by the formula:

$$n_h = \frac{N_h}{N} n \quad (2)$$

where:

n_h = number of sample from each village

Based on the formula (2) the number of samples at Sejahtera Village totaled 48 cocoa farmers and Bulili Village totaled 50 cocoa farmers.

Data Analysis Techniques

Data analysis techniques which were used in this research as follows:

1. to answer the first objective was used of descriptive statistics and
2. to answer the second objective was used multiple regression analysis with the model:

$$Y = \beta_0 + \beta_1 X_1 + \beta_2 X_2 + \beta_3 X_3 + \beta_4 X_4 + \beta_5 X_5 + \beta_6 X_6 + \mu \quad (3)$$

Where:

Y = level of technology adoption side-grafting (%)

β_0 = intercept

β_1 - β_6 = regression coefficients

X_1 = dummy ethnic

D = 0 for ethnic of Kaili (indigenous people)

D = 1 for non-ethnic Kaili (immigrants)

X_2 = age (year)

X_3 = education (year)

X_4 = cocoa farming experience (year)

X_5 = number of household members (people)

X_6 = frequency follow agricultural extension (frequency)

μ = error term

RESULTS AND DISCUSSION

Level of Technology Adoption Side-Grafting

Level of technology adoption side-grafting was the ratio of the cocoa farming area was side-grafting with of the cocoa farming area which was owned by farmers. Level of technology adoption side-grafting at Sigi regency-Indonesia were shown in Table 1.

Table 1: Level of Technology Adoption Side-Grafting on Cocoa Farming at Sigi Regency-Indonesia

Level of Technology Adoption Side-Grafting (%)	Frequency (people)	Percentage (%)
0	55	56,12
50	3	3,06
60	1	1,02
70	3	3,06
75	1	1,02
80	5	5,10
100	30	30,61
Total	98	100,00

Table 1 showed there were 56.12% of the farmers did not perform side-grafting on cocoa farming and 30.61% of farmers who have done side-grafting to all the cocoa which was owned. This means that the level of technology adoption side-grafting on Cocoa Farming was still low at Sigi regency-Indonesia. Side-grafting on cocoa farming was to paste entris (branch plagiotrop) derived from types (clones) of superior cacao on rootstock cacao plants that old or have low productivity. Shoots that grew from the scion then developed into trunks and branches, which in turn produced fruit with higher productivity level according to the potential of genetic material cropping (Basri, 2009). Some researchers showed that side-grafting could increase the production and productivity of cocoa trees (Limbongan dan Langsa, 2006; Muis dan Basri, 2008).

FACTORS AFFECTING LEVEL OF TECHNOLOGY ADOPTION SIDE-GRAFTING

Multiple regression analysis was used to analyze the factors that affected the level of technology adoption side-grafting on cocoa farming. Level of technology adoption side-grafting on cocoa farming could be affected by the characteristics of farmers, such as of ethnic, age of farmers, farmer education, experience of cocoa farming, number of household members, and frequency of follow agricultural extension. Results of multiple regression analysis were shown in Table 2. Table 2 showed Fstatistic = 43.119 with probability 0.000 < 0.010 (α = 1%) proved reject the null hypothesis, meant that the independent variable ethnic, age of farmers, farmer education, experience of cocoa farming, number of household members, and frequency of follow agricultural extension of affect simultaneous on the level of technology adoption side-grafting on cocoa farming. Determinant coefficient of adjusted R square = 0.7226 showed that variation the level of technology adoption side-grafting on cocoa farming could be explained by the independent variables ethnic, age of farmers, farmer education, experience of cocoa farming, number of household members, and frequency followed agricultural extension as big as 72.26%, while 27.74% was explained by other factors not included in the model. Effect of each factor to the technology adoption side-grafting on cocoa farms were as follows:

Table 2: Results of Parameter Estimation Equation Level of Technology Adoption Side-Grafting on Cocoa Farming

Variable	Coefficients	Standard Error	probability	VIF
Intercept	-0.381	15.173		
Ethnic	20.679	6.041	0.001**	1.468
Age	-0.918	0.273	0.001**	1.249
Education	1.738	4.010	0.666	2.994
Experience of cocoa fanning	3.217	0.642	0.000**	2.761
Number of household members	1.752	1.701	0.305	1.114
Frequency following the agricultural extension	2.865	1.124	0.013*	2.815
Adjusted R Square 0.7226				
F statistic 43.119**				

** = Significant at $\alpha = 1\%$, * = significant at $\alpha = 5\%$.

Ethnic

Ethnic of significant effect on the level of technology adoption side-grafting on cocoa farming at Sigi regency-Indonesia at $\alpha = 1\%$. Coefficient of 20.679 meant that, immigrant farmers (non-ethnic kaili) had a level of technology adoption side-grafting on the cocoa farming as big as 20.679% higher than indigenous farmers (ethnic Kaili). This was due to immigrant farmers (Bugis, Toraja, Java, Masamba, Batak, Seko, Sanger, and Palopo) more enthusiastic to live more advanced because it was in the overseas region. They more readily accepted innovations to increase welfare. This research relefan with of research Jasmine, Norhamidah, Suwaiba, and Nordin (2012) on the factors associated with continuance adoption of agricultural innovations among rural people at Sabah, Malaysia. This qualitative research conducted in rural areas at the City of Marudu, Sabah, Malaysia. The findings of the research showed that five major factors associated with continuance adoption of agricultural innovations were the demographic and socio-economic, psychological and cognitive attributes of leadership, communication, provision of inputs and technical assistance.

Age

Age of farmers significant effected on the level of technology adoption side-grafting on the cocoa farming $\alpha = 1\%$. Coefficient of -0.918 meant that every increase in age farmers as big as 1 year could reduce the level of technology adoption side-grafting on cocoa farming as big as 0.918%, with assume other factors held constant. This was due to the older increasingly farmers increasingly retain their mores. Farmers of age young more eager found new things, so they were more quickly its adopt. This research was supported by research Paxton et al. (2011), through a 2005 survey of cotton producers in the southeastern United States. Results of this research showed that age was negatively correlated with the adoption of agricultural technologies. These results showed that young age producers were higher adopt of agricultural technology. The same thing was stated by Moga, Constantin, and Antohi (2012) elderly could lead to a negative attitude towards the use of Information and Communication Technology in Romania. However, contrary to research Xu and Wang (2012), Lestrelin et al. (2011) and Singha, Baruah, Bordoloi, Dutta, and Saikia (2012) showed the adoption did not depend on the variable age of farmers.

Farmer Education

Farmer education did not significant effect on the level of technology adoption side-grafting on cocoa farming. This was due largely less educated respondents, so it did not significant affect of the technology adoption side-grafting. This research was supported by research Lestrelin et al. (2011) that an evaluation study in 21 villages in Laos, the technology adoption of conservation agricultural (CA) had become an important constituent of the agricultural landscape. Statistical and qualitative evidence showed that experimentation and adoption did not depend on education. The same was stated by Raut, Sitaula, Vatn, and Paudel (2011) that education did not affect the farmer's decision about the application of intensive agriculture at Nepal. This research contrary with research of Xu and Wang (2012), that education effected technology adoption decisions by fruit growers in China. Research Singha et al. (2012) at Sonitpur Assam District, India, showed that the education variables had a significant positive correlation with the adoption of technology. The same thing was said by Abdullah and Samah (2013) that the level of education affected of technology adoption among farmers in Malaysia.

Experience of Cocoa Farming

Experience of cocoa farming significant effected on the level of technology adoption side-grafting on the cocoa farming $\alpha = 1\%$. Coefficient of 3.217 meant that every increase in experience of cocoa farming as big as 1 year could increase the level of technology adoption side-grafting on cocoa farming as big as 3.217%, with assumed other factors held constant. This was due to the increasingly higher of experience a farmer, the more extensive their knowledge on innovation and technology, so they were more open to new technologies. This research was supported by research Ayoola (2012) who studied the effect of socio-economic factors to the yams technology at Nigerian. Probit analysis showed that experience farming significant effected on technology adoption. The same was stated by Nyanga (2012) that the adoption of conservation agriculture technologies (CA) was quite low in most of Africa. However, Zambia had been quite successful in increasing the adoption of CA by smallholder farmers. This research had shown that previous experiences significantly increase adoption of CA.

Number of Household Members

Number of household members did not significant effect on the level of technology adoption side-grafting on cocoa farming. This was due to the number of household members mostly had not been productive, so the number of household members did not affect the level of technology adoption side-grafting. Research Bello, Salau, and Ezra (2012) that identified the factors that affected of rice technology discontinuation at Nasarawa, central Nigeria. Household size accounted for 67.0% of the variation in the termination of rice technology adoption by farmers.

Frequency Followed the Agricultural Extension

Frequency followed the agricultural extension of significant effected on the level of technology adoption side-grafting on the cocoa farming $\alpha = 5\%$. Coefficient of 2.865 meant that any increase in the frequency followed the agricultural extension as big as 1 time could increase the level of technology adoption side-grafting on cocoa farming as big as 2.865%, with assume other factors held constant. This was due to the increasingly higher of frequency followed the agricultural extension increasingly the more their knowledge of innovation and technology, so they were more open to new technologies. This research was supported by research Ayoola (2012) who studied the effect of socio-economic factors to the yams technology at the Nigerian. Probit analysis showed that extension frequency significant effected to the technology adoption. Further Bello et al. (2012) identified the factors that affected of rice technology discontinuation at Nasarawa, central Nigeria. Characteristics of farmers such as extension contact accounted for 67.0% of the variation in the termination of rice technology adoption by farmers.

CONCLUSION AND RECOMMENDATIONS

The results showed the level of technology adoption side-grafting on cocoa farming was still low at Sigi regency-Indonesia. Characteristics of farmers such as ethnicity, age, cocoa farming experience, and frequency following agricultural extension of significant effect on the level of technology adoption side-grafting on cocoa farming at Sigi regency-Indonesia. Therefore improving human resources in rural areas would increase of the technology adoption side-grafting on cocoa farming in the Sidi Regency-Indonesia.

REFERENCES

1. Abdullah, F. A., & Samah, B. A. (2013). Factors Impinging Farmers' Use of Agriculture Technology. Asian Social Science, 9(3), 120-124. http://dx.doi.org/10.5539/ass.v9n3p120

2. Anonimous. (2012). Kebutuhan Kakao Dalam Negeri 280.000 Ton. Retrieved 11 Nopember, 2012, from http://id.berita.yahoo.com/kebutuhan-kakao-dalam-negeri-280-000-ton-154412488.html
3. Ayoola, J. B. (2012). Socio-economic Determinants of the Adoption of Yam Minisett Technology in the Middle Belt Region of Nigeria. Journal of Agricultural Science, 4(6), 215-222. http://dx.doi.org/10.5539/jas.v4n6p215
4. Badan Pusat Statistik. (2012). Kabupaten Sigi Dalam Angka. Kantor Pusat Statistik Kabupaten Sigi, Sulawesi Tengah. Basri, Z. (2009). Kajian Metode Perbanyakan Klonal pada Tanaman Kakao. Media Litbang Sulawesi Tengah, 2(1), 7-14.
5. Bello, M., Salau, E. S., & Ezra, L. (2012). Analysis of Factors Influencing iscontinuance of Technology Adoption: The Situation with Some Nigerian Farmers. Sustainable Agriculture Research, 1(2), 292-300. http://dx.doi.org/10.5539/sar.v1n2p292
6. Departemen Pertanian Direktorat Jenderal Perkebunan. (2009). Buku Panduan Teknis Budidaya Tanaman Kakao (Theobroma cacao L.). Gerakan Peningkatan Produksi dan Mutu Kakao Nasional.
7. Direktorat Jenderal Perkebunan. (2012). Buku Statistik Perkebunan Tahun 2009-2011.Retrievedfrom,http://www.google.co.id/#bav=on.2,or.r_qf.&fp=e0d9043041582470&q=Buku%20Statistik%20Perkebunan%20kakao%20Tahun%202009%20%E2%80%93%202011%20pdf, diakses 11 Nopember 2012
8. Jasmine, J. H., Norhamidah, A. M., Suwaiba, J. Z., & Nordin, M. (2012). Factors Associated with the Continuality of Agricultural Innovation Adoption in Sabah, Malaysia. Journal of Sustainable Development, 5(1), 47-54. http://dx.doi.org/10.5539/jsd.v5n1p47
9. Lestrelin, G., Hoa, T. Q., Jullien, F., Rattanatray, B., Khamxaykhay, C., & Tivet, F. (2011). Conservation agriculture in Laos: Diffusion and determinants for adoption of direct seeding mulch-based cropping systems in smallholder agriculture. Renewable Agriculture and Food Systems, 27(1), 81-92. http://dx.doi.org/10.1017/S174217051100055X
10. Limbongan, J., & dan Langsa, Y. (2006). Peremajaan pertanaman kakao dengan klon unggul melalui teknik sambung samping (side-

cleft grafting) di Sulawesi Tengah. Prosiding Seminar Nasional Pengembangan Usaha Agribisnis Industri Pedesaan, Palu.

11. Moga, L. M., Constantin, D. L., & Antohi, V. M. (2012). A Regional Approach of the Information Technology Adoption in the Romanian Agricultural Farms. Informatica Economică, 16(4), 29-36.

12. Muis, A., & dan Basri, Z. (2008). Kajian Peningkatan Produksi dan Pendapatan Usahatani Kakao Melalui Teknik Sambung Samping. Media Litbang Sulawesi Tengah, 1(2), 78-87.

13. Neilson, J. (2008). Program Gerakan Nasional Percepatan Revitalisasi Kakao Nasional (GERNAS). Masukan strategis dari Forum Kemitraan Kakao Berkelanjutan (Cocoa Sustainability Partnership), ACIAR. Retrieved 11 Nopember, 2012, from

14. http://www.google.co.id/#bav=on.2,or.r_qf.&fp=442bdad5c6096e1d&q=Program+Gerakan+Nasional+Percepatan+Revitalisasi+Kakao+Nasional+%28GERNAS%29

15. Nyanga, P. H. (2012). Factors Influencing Adoption and Area under Conservation Agriculture: A Mixed Methods Approach. Sustainable Agriculture Research, 1(2), 27-40. http://dx.doi.org/10.5539/sar.v1n2p27

16. Parel, C. P., Caldito, G. C., Ferrer, P. L., De Guzman, G. G., Sinsioco, C. S., & Tan, R. H. (1973). Sampling Design and Procedures. The Agricultural Development Council. Quezon City.

17. Paxton, K. W., Mishra, A. K., Chintawar, S., Roberts, R. K., Larson, J. A., English, B. C., ... Martin, S. W.(2011). Intensity of Precision Agriculture Technology Adoption by Cotton Producers. Agricultural and Resource Economics Review, 40(1), 133-144.

18. Raut, N., Sitaula, B. K., Vatn, A., & Paudel, G. S. (2011). Determinants of Adoption and Extent of Agricultural Intensification in the Central Mid-hills of Nepal. Journal of Sustainable Development, 4(4), 47-60. doi:10.5539/jsd.v4n4p47

19. Singha, A. K., Baruah, M. J., Bordoloi, R., Dutta, P., & Saikia, U. S. (2012). Analysis on Influencing Factors of Technology Adoption of Different Land Based Enterprises of Farmers under Diversified Farming System. Journal of Agricultural Science, 4(2), 139-146. http://dx.doi.org/10.5539/jas.v4n2p139

Chapter 10

Color Characteristics of Cocoa Butter Made From the Processing of the Ivorian Cocoa Beans

David Koffi Akaki[1*], Eric Fattoh Elleingand[1], Sadat AW[1], and Coulibaly Kelegoun[1]

[1]Laboratoire des Procédés Industriels de Synthèse et de l'Environnement et des Energies Nouvelles (LAPISEN). Département de Génie Chimique et Agroalimentaire (GCAA) de L'Institut National Polytechnique Félix Houphouët Boigny (INP-HB) BP 1313 Yamoussoukro, République de CÔTE D'IVOIRE.

*Corresponding Author: David Koffi AKAKI

ABSTRACT

Color of cocoa butters was measured using a CM-2002 Minolta Spectrocolorimeter. The results showed that the "L" values (lightness) varied from 73.11 to 84.76, the "a" values (greenness) varied from –2.15 to –1.08 and a redness of 4.99, and the "b" values (yellowness) varied from 28.63 to 41.11. Acidities of all types cocoa butters varied from 0.47 to 10.26. The best cocoa butters were obtained from staly and fully purple beans and cocoa beans of grades 4, 5, and 6. These butters had acidities of 0.48 and 0.56 for fully purple and staly beans and 2.41, 1.30, and 0.47 for grades 4, 5, and 6, respectively.

INTRODUCTION

Color is an important factor to both consumers and producers in most foods including cocoa butter. This factor is appreciated through many color measurement systems. Among these systems, the Hunter L*a*b system is one of the most used in the food industry. The Hunter color measurement method is based upon the following parameters which are lightness (L), greenness (-a), redness (a), yellowness (b), and blueness (- b) (Deman 1980). All these parameters measured with a colorimeter are greatly affected by fermentation. Color of cocoa butter depends upon factors such as fermentation and fermentation time, acidity, drying, and all the steps of the processing.

Fermentation is one of the essential steps in the processing of cocoa beans. During this step, structural and chemical changes occur as the result of enzymatic activities. Organic acids like acetic and lactic acids are produced (Jinap 1994). A recent study has reported that fermentation was responsible for the reduction of protein content down to 57% (Lerceteau and others 1999). Jinap and others (1994) showed that methyl pyrazine and many of its related chemical compounds were formed during fermentation of cocoa beans in Malaysia.

Bacillus subtilis and B. megaterium were the bacteria responsible for the fermentation. In addition to these factors, the color of cocoa butter depends largely on the quality of cocoa beans. Upon harvesting and fermentation, cocoa beans can be normal, fully purple, molded, or staly. Cocoa butters obtained from these different types of beans vary according to the quality of the beans. Cocoa butter can also be obtained from the shell and such a butter has some typical characteristics.

A number of studies have also related the quality of cocoa butter to acidity. Hashim and others (1997) showed that the oxidation of cocoa butter was induced through heating to 90 °C. The cocoa butter could be stabilized by using alpha-tocopherol. In a similar investigation, acetic and lactic acids were shown to have a tremendous effect on cocoa beans and probably on the cocoa butters obtained from the processing of the corresponding beans (Selamat 1998).

Although many studies have already been focussed on cocoa butter, up to day, there still is no information available on the characteristics of the ivorian cocoa butter. Therefore, this investigation was carried out, first of all to study the effect of the quality of cocoa beans on the color of cocoa butter and second to study the impact of mixing different types of beans on the color of the corresponding cocoa butters.

MATERIALS AND METHODS

Methods of Fermentation

Cocoa beans previously fermented and dried were commercially received. Beans were naturally fermented with wild microflora throughout cocoa plantations across the country. The fermentation was carried out into holes dug on the ground. Holes were dug on elevated ground in a tilted manner in order to easily collect the sweating cocoa liquid during the fermentation process (Buamah and others 1997). After digging the hole, its bottom and sides were covered with large banana leaves. Then holes were filled with about one hundred kilograms (kg) of cocoa beans. Holes filled with beans were covered with other banana leaves. The fermentation was carried out over seven (7) to nine (9) days. Beans were moved to new holes after every two (2) days of fermentation. Another method of fermenting beans was to use perforated plastic bags. Bags were filled with 50 to 60 kg of beans and hung on the branches of some trees. Wooden supports were also made to hang the bags filled with the beans. Bags were replaced every forty eight (48) hours as previously mentioned.

Following fermentation, the beans were removed and sun dried on cemented flat surfaces. Then the fermented and dried beans were sold to cocoa processing plants such as "la société africaine de cacao" (SACO). Beans were separated into normal, molded, staly, and fully purple and processed according to the scheme outlined in figure 1. Upon fermentation, beans are either normal or faulty. Faulty beans are fully purple, molded, or staly. Fully purple beans are cocoa beans that are insufficiently fermented. Anthocyanins content in these beans is high. Tannins also have not undergone enzymatic reaction to degrade compounds responsible for the bitter taste. Beans are then fully purple and contain all colorants of unfermented cocoa beans. Staly cocoa beans are ripened beans that have not been fermented. They are characterized by a bitter taste and a high level of anthocyanins. Tannins in these beans have not undergone enzymatic reaction and the beans have not been colored in brown. Chemical compounds formed during fermentation do not exist and the beans exhibit an offflavor.

Molded cocoa beans are formed following growth of molds. Then lipolytic enzymes derived from the molds attack cocoa butters and lead to off-flavor. A fifth type of butter was obtained from the processing of the cocoa shell. Ten (10) Kg of each type of bean were weighed and processed according to the scheme of figure 1. The color parameters were measured using the CM-2002 Minolta type spectrocolorimeter with the following operating conditions:

Lightness: D 65

Geometry measurement: d / s

Standard Observer CIE: 10° - SCI

Measurement on a surface of 8 mm

The mean of three values is recorded

Colorimeter system: L, a, and b.

Composition of Grades

Cocoa beans commercially obtained were mixed in the following proportions to come up with six (6) grades:

Grade 1: 3 % molded, 3 % staly, 3 % purple, and 91 % normal beans

Grade 2: 4 % molded, 8 % staly, 8 % purple, and 80 % normal beans

Grade 3: 10 % molded, 8 % staly, 8 % purple, and 74 % normal beans

Grade 4: 25 % molded, 25 % staly, 25 % purple, and 25 % normal beans

Grade 5: 10 % molded, 50 % staly, 25 % purple, and 15 % normal beans

Grade 6: 0 % molded, 50 % staly, 50 % purple, and 0 % normal beans

Each type or lot of cocoa beans was processed to obtain a cocoa butter (figure 1). Then the color parameters which are lightness (L), greenness (a), and yellowness (b) were measured using the Minolta CM-2002 spectrocolorimeter with the operating conditions previously stated.

Sample Preparation

From each cocoa butter obtained after processing of each type of beans, thirty (30) ml of the liquid cocoa butter were poured into a 50 ml beaker and the beaker was placed in a cold room at 6 to 7 °C for 10 to 15 min. As the butter started to solidify, the liquid and solid phases were thoroughly mixed and transferred onto a tray. The tray was again placed in the cold room for 15 to 20 min. Then, the tray was removed from the room, and the butter was taken off the tray, and the color factors (L, a, and b) were rapidly measured on the surface.

Each color parameter was measured through five (5) replicates at each stage of the processing and the mean, standard deviation, and the coefficient of variation were computed. The analysis of variance (ANOVA) was performed by the SAS procedure (SAS Institute, Inc. 1985). Like color, the acidity of each cocoa butter was determined according to the SACO method of acidity titration. Five (5) grams of each cocoa butter were weighed out. Then fifty (50) ml of an ether-ethanol (50:50) solution was added to the liquid butter. The mixture was titrated with a 0.1N potassium hydroxide solution. Acidity was calculated according to the following formula:

$$\text{Acidity} = (2 \times 5 \times V) : (P \times N) \quad (1)$$

V: Volume of the potassium hydroxide solution used to neutralize the acids.

P: Weight of the sample (cocoa butter)

N: Normality of the potassium hydroxide solution (0.1N)

The acidity is expressed as the quantity of the oleic acid

Five (5) measurements were performed on each sample and the mean was computed.

Differences among means for the color parameters and acidities of the butters obtained from the five qualities of beans were determined by the Fischer's least significant difference (LSD) with $p < 0.05$.

RESULTS AND DISCUSSION

Figure 1 outline the processing scheme of cocoa beans. First of all, beans were fermented to develop the desired flavor. This was achieved through production of acids such as acetic and lactic acids; pyrazines, aldehydes, and peptides. The effect of fermentation on the chemical properties of cocoa beans as well as cocoa butter has been extensively reported (Jinap and others 1994; Brito and others 2001; Schwan and others 1995).

Fermentation is an essential and unique operation applied to cocoa beans that provide cocoa butter with a desirable flavor and an acceptable acidity. Brito and others (2001) studied the chemical changes induced by fermentation, drying, and roasting on cocoa beans. These authors showed that fermentation up to three (3) days increased the amount of free amino acids in cocoa beans. Jinap and others (1994) showed that the acidity and volatile fatty acids content were related to the step of drying. Schwan and others (1995) reviewed fermentation of cocoa beans by yeasts, lactic and acetic bacteria, aerobic spore forming bacteria and fungi. This review was mainly focussed on the enzymatic degradation of the pulp of the cocoa beans. Mabett (1987) reported that fermentation and drying were the most critical steps of cocoa processing as they depend upon biological systems difficult to control. The procedure known as "curing" starts from the dehusking of wet beans and ends by the production of dried beans ready to be stored or shipped.

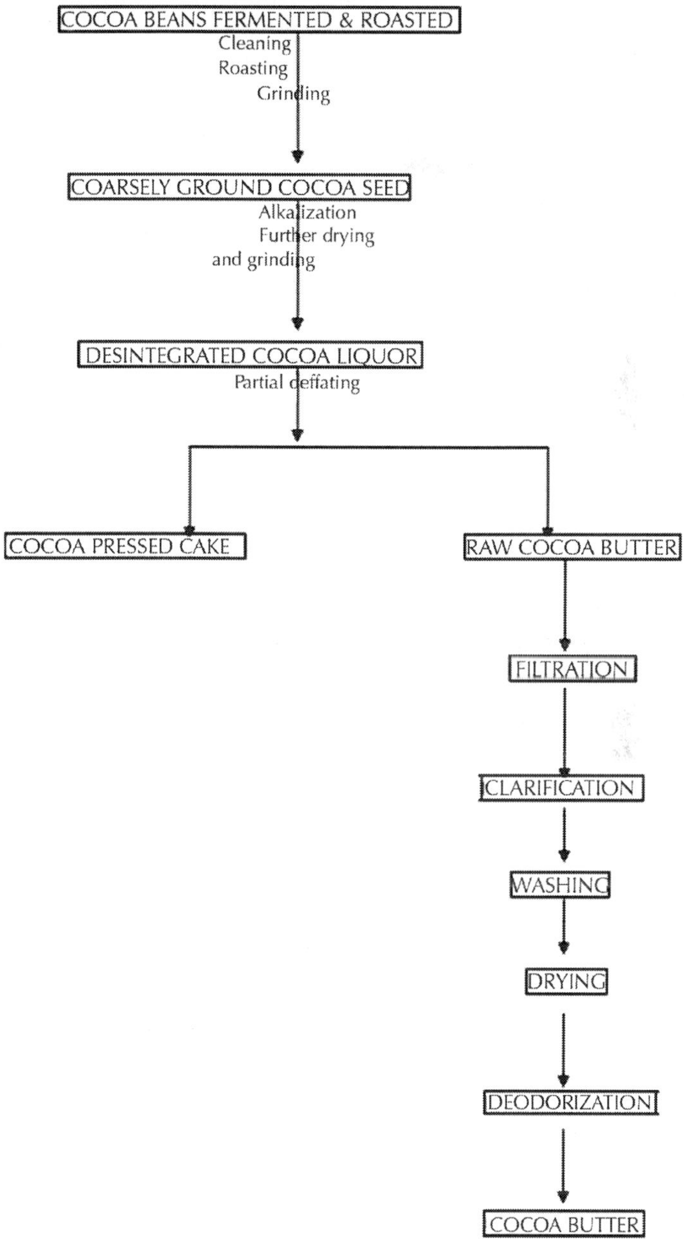

Figure 1: Processing of fermented and roasted cocoa beans to produce cocoa butter.

Results in table 1 shows the measurements of the color factors which are lightness (L), greenness (-a), redness (a), and yellowness (b) of cocoa butters obtained from the processing of the different types of cocoa beans.

Table 1: Measurements the three color factors lightness (L), greenness (- a), redness (+ a), and yellowness (b) of the butter obtained from the shell of the cocoa beans, the normal, molded, fully purple; and staly cocoa beans

Color			**Types of beans**							
	Normal˙		Molded		Staly˙		Fully purple·		Cocoa shell[s]	
	Mean	CV	Mean	CV	Mean	CV	Mean	CV	Mean	CV
Lightness (L)	82.30′	0.17	80.89[b]	0.16	84.50′	0.34	83.89[d]	0.34	73.11′	0.11
Greenness (- a)	-1.08′	7.40	- 1.21[b]	4.13	- 2.15`	5.58	- 2.14′	3.73	4.99 [d]	1.00
Redness (+ a)										
Yellowness (b)	30.89[a]	2 .29	33.24′	1.44	30.81′	0 .65	29.02′	1.58	41.11[d]	0.46

CV: coefficient of variation

*All values are the means of five measurements.

[a - e]Means in the same row with the same letter are not significantly different ($p < 0.05$)

Among the five types of cocoa butters, the value of lightness varied from 73.11 (butter obtained from the shell) to 84.50 (cocoa butter made from staly beans). The staly beans yielded the cocoa butter with the highest brightness while cocoa butter obtained from the shell of the beans gave a cocoa butter with the lowest clarity. Cocoa butter obtained from the shell of cocoa beans exhibits a red color appearance with a strong yellowness and less lightness than the other butters. Such a color for cocoa butter is undesirable (Saré 2001). In addition, all five types of cocoa beans yielded cocoa butters with significant different brightness. While measuring lightness, the coefficients of variation for all five raw materials were less than 1%. Therefore, the spectrocolorimeter was consistent in measuring lightness of the cocoa butter.

Values of greenness varied from – 2.15 (staly beans) to 4.99 (shell type beans) with CV of 1 to 7%. Except the cocoa butter obtained from the shell, which had red appearance, all four other types of beans yielded a cocoa butter with some green color. Also, the staly and fully purple beans provided the cocoa butters with the greenest color. However, the CV were slightly higher than those obtained while measuring lightness. Cocoa butters obtained from processing of staly and purple beans of "a" values (greenness) of – 2.15 and – 2.14, respectively, did not show any significant difference. Though, the other three types of butter were significantly different in redness and greenness. The highest CV was 7.40% (normal beans) and the lowest was 1% (shell).

It appears that there was some slight variation among the values of greenness and redness of the cocoa butter. As far as yellowness was concerned, the "b" values ranged from 29.02 for cocoa butter of fully purple beans to 41.11 for the butter obtained from the shell. Coefficients of variation varied from 0.46% (shell) to 2.29% (normal beans). Like lightness and greenness, yellowness for the butter obtained from fully purple beans (b = 29.02) was close to that of staly beans (b = 30.81). Moreover, the degree of yellowness of the butter obtained from normal beans (b = 30.89) was statistically the same with that of staly beans (b = 30.81). While measuring the intensity of yellowness, the CV were less than 3% for all types of butters. Therefore, the spectrocolorimeter was repeatable in measuring the three color parameters. Consistency of the CM-2002 Minolta spectrocolorimeter has been recently noted (Saré 2001).

Rigel and others (1997) have shown a significant difference between butters made from a variety of beans. These authors reported that saturated fatty acids in cocoa butter were predominant as compared to unsaturated fatty acids. Among these acids, palmitic and stearic acids were the most abundant.

Results in table 2 represent the acidity of the cocoa butters obtained with the five types of beans. Acidity varied from 0.48 (cocoa butter of purple beans) to 10.26 (cocoa butter of molded beans). Coefficients of variation were all less than 3%. The highest acidity of 10.26 was observed with the cocoa butter obtained from the molded beans. Acidity represents the percentage of free fatty acids that are mainly palmitic, stearic, and oleic acids. These results show that the best cocoa butters were obtained from the processing of purple and staly beans. The next better cocoa butter was the one obtained from the processing of the normal beans.

Table 2: Acidity of the cocoa butter obtained from the processing of the different types of beans

Acidity		
Cocoa butter	Mean	CV
Normal beans	1.47.	0.68
Molded beans	10.26[6]	0.00
Staly beans	0.59	1.78
Fully purple beans	0.48′	2.08
Cocoa shell	8.45[d]	0,00

All values are the mean of five measurements

CV: Coefficient of variation

[a-d]Means in the same column with the same letter are not significantly different ($P < 0.05$).

There seems to be a correlation between lightness and the percentage of acidity of the different cocoa butters. Formation of purple beans has been reported to be related to the method of drying. A research focused on the comparison of four methods of drying which were indirect solar drying, direct solar drying, open air with a perforated steel surface, and open air with a non perforated wooden surface. These four methods of drying were examined at three loading rates of 13.7, 26.9, and 40.4 kg/m^2. Beans dried at the rate of 13.7 kg/m^2 had the best color with a high titrable acidity. Beans that were dried at the rate of 40.4 kg/m^2 showed a significant low acidity with a poor color. This study concluded that the beans dried indirectly exhibited the best quality.

Although, staly and purple beans gave significant different cocoa butters, values of lightness were higher than those of the other butters. The worst cocoa butter with less lightness appears to be the butter extracted from the shells of the cocoa beans.

Table 3: Acidity of the cocoa butter obtained from the processing of the six grades of cocoa beans

Acidity		
Cocoa butter grade	Mean	CV

I	1.35′	0.00
2	I.46a	0.00
3	I.90⁶	0.00
4	2.41′	0.00
5	1.30′	0.77
6	0A7″	2.12

All values are the mean of five measurements

CV: Coefficient of variation

[a-d]Means in the same column with the same letter are not significantly different (P < 0.05).

The acidity (titrable acidity) of the six grades is shown in table 3. The lowest acidity was obtained with the sixth grade composed of 50 % of staly beans and 50 % of fully purple beans. The other five grades have their acidities varying from 1.30 (grade 5) to 2.41 (grade 4). Overall, the acidity was low and therefore acceptable. These low acidities imply that there were less free fatty acids present in all grades of beans processed to obtain the butter. Since free fatty acids are usually yielded upon oxidation, we therefore believe that lipid oxidation was very little in all cocoa butter.

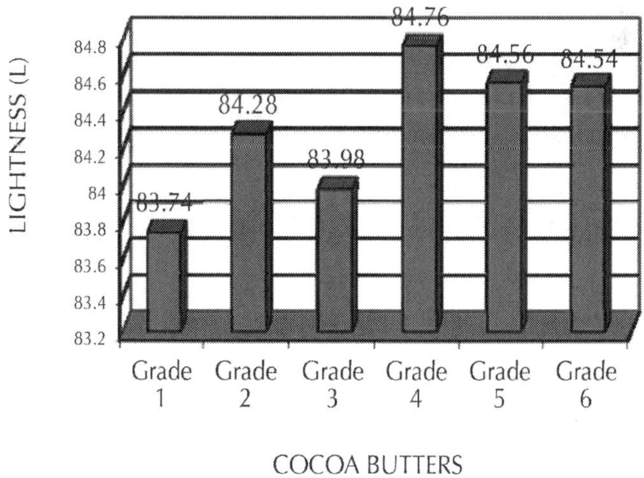

Figure 2: Measurements of the lightness (L) of the cocoa butters obtained from the processing of the different types of grades.

The color of the cocoa butters obtained from the different grades is represented from figure 2 through figure 4. Figure 2 shows the graphic comparison of lightness of the six types of butters. Lightness of all grades of cocoa butters varied from ("L" value) 83.74 to 84.76. There was no significant difference in lightness between cocoa butters of grade 1 and that of grade 2; between grades 2 and 3; and between grades 4, 5, and 6. The highest lightness was observed with the butters of grades 4, 5, and 6. A closer analysis of the composition of the beans of each grade shows that lightness might be related to staly and fully purple beans.

Normal beans with proportions of 91 and 80% in grades 1 and 2, respectively yielded cocoa butters with significantly lower lightness than that of grades 4, 5, and 6. Figure 3 shows the graphic representation of greenness of the six types of cocoa butters. The values of greenness varied from – 2.05 to – 1.38. Cocoa butters obtained from beans of grades 2 and 3 were not significantly different. Except grades 2 and 3, all the others beans gave cocoa butters with significantly different greenness. The lowest greenness was observed with the butter obtained from beans of grade 5 while the highest greenness derived from grades 2 and 3. In the previous analysis, we noticed that beans of grade 5 yielded the cocoa butter with the highest lightness.

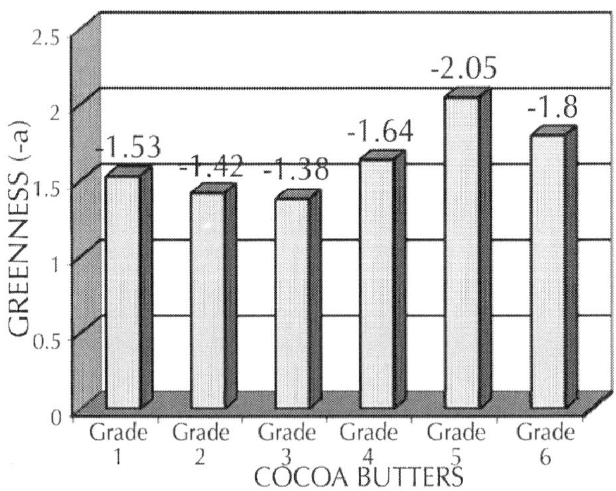

Figure 3: Measurements of the greenness (- a) of the cocoa butters obtained from the processing of the different types of grades.

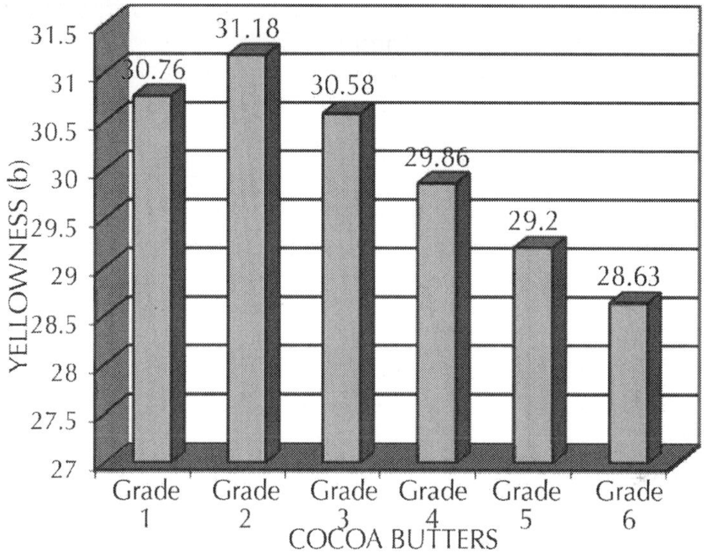

Figure 4: Measurements of the yellowness (b) of the cocoa butters obtained from the processing of the different types of grades.

As far as yellowness was concerned, the "b" values in figure 4 varied from 28.63 (cocoa butter of grade 6) to 31.18 (cocoa butter of grade 2). Cocoa butters of beans of grades 1 and 3 were statistically the same.

Except the butters of these two grades, all the other butters were significantly different in yellowness. The butter obtained from the processing of the beans of grade 2 had the highest intensity of yellowness. Unlike lightness and greenness, we did not notice the highest yellowness with the butters obtained from the processing of cocoa beans of grades 4, 5, and 6.

CONCLUSIONS

Among the different types of cocoa beans, the best cocoa butters were the ones obtained from the processing of the staly and fully purple beans (tables 1 and 2). These butters had "L" values (lightness) of 84.50 and 83.89 for staly and fully purple beans, respectively. The next brightest cocoa butter was the one obtained from normal beans with an "L" value of 82.30. The worst cocoa butter in lightness with an "L" value of 73.11 was the one obtained from the shell. All cocoa butters had some greenness except the one extracted from the shell that exhibited some redness with an "a" value of 4.99. Greenness was statistically the same for the butters yielded by staly and fully purple beans. As far as yellowness was concerned, there was no significant difference between butters obtained from staly and normal beans. Butter obtained from processing of the shell showed the highest significant yellow color (b = 41.11).

Influence of acidity on cocoa butters' quality has been investigated. Holm and others (1993) studied the effects of the organic acids in cocoa on the flavor of chocolate. Their study reported the presence of acids such as oxalic acid in cocoa based products. The research showed that oxalic acid was responsible for flavor characteristics in chocolate. Another investigation showed that 40% of ripened cocoa beans contributed to the color of chocolate. The proportion of these beans might also well play an important role in the color development of cocoa butter (Samah and others 1993).

The highest acidity was observed with the butter obtained from molded beans (acidity of 10.26 (table 2)). The next highest acidity of 8.45 was observed with the butter processed from the shell. Butters with the lowest acidity were seen with purple (0.48) and staly (0.56) beans. Overall, there was a relation between color and acidity of the cocoa butter. Butters with a nice light and yellowish appearance had the lowest acidity.

These butters were yielded by the fully purple, staly, and normal cocoa beans. The results shown in figures 2, 3, and 4 correlate well with the effect of beans' quality on the overall color of cocoa butter. These data indicate that cocoa butters of highest lightness are obtained with cocoa beans of grades 4, 5, and 6. These three grades are composed of 25% of molded, staly, fully purple, and normal beans for grade 4; 10% of molded, 50% of staly, 25% of fully purple, and 15% of normal beans for grade 4; and 50% of staly and fully purple beans. It appears that the beans that contribute best to the color of the best cocoa butters are the staly and the fully purple beans. In previous studies, molded beans were shown to be responsible of a cocoa butter of high acidity. The highest greenness (b = - 2.05) was also observed with cocoa butter obtained from processing of beans of grade 5. Though, the cocoa butter with the most important intensity of yellowness was observed with beans of grade 2. The best grades of cocoa beans which are grades 4, 5, and 6 yielded cocoa butters that had less significant yellowness than those of cocoa butters obtained from grades 1, 2, and 3. Overall, results of this investigation show that color of cocoa butters largely depends upon the state of the cocoa beans.

REFERENCES

1. Buamah R, Dzogbefia VP, Oldham JH. 1997. Pure yeast culture fermentation of cocoa (Theobroma cacao L): effect on yield of sweatings and cocoa bean quality. World Journal of Microbiology & Biotechnology, 13 (4):457-462.
2. Brito ES-de, Pezoa-Garcia NH, Gallao MI, Cortelazzo AL, Fevereiro PS, Braga MR. 2001. Structural and chemical changes in cocoa (Theobroma cacao L) during fermentation, drying, and roasting. J. Sci. Food Agric., 81(2):281-288.
3. Deman JM. 1980. Ch 6. In "Principles of Food Chemistry", Van Nostrand Reinhold Company, Inc., New York, New York, p 189-226.
4. Hashim L, Hudiyono S, Chaveron H. 1997. Volatile compounds of oxidized cocoa butter. Food Research International, 30(3,4):163-169.
5. Holm CS, Aston JW, Douglas K. 1993. The effects of the organic acids in cocoa on the flavour of chocolate. J. Sci. Food Agric., 61(1):65-71.
6. Jinap S 1994. Organic acids in cocoa beans. Asean Food Journal, 9(1) :3-12.
7. Jinap S, Siti MH, Norsiati MG, Selamat J. 1994. Formation of methyl pyrazine during cocoa bean fermentation. Pertanika Journal of Tropical Agriculture Science, 17(1) :27-32.
8. Lerceteau, E, Rogers J, Petiard V, Crouzillat D. 1999. Evolution of cacao bean proteins during fermentation : a study by twodimensional electrophoresis. J.Sci. Food Agric., 79(4):619-625.

9. Mabett T. 1987. Fermentation and drying : keys to cocoa quality. Coffee and Cocoa International, 14(5):59-60.
10. Rigel L, Fanny-C P, Agricia Q. 1997. Characteristics of cocoa butter extracted from Criollo cultivars of the Theobroma cacao L. Food Research International, 30(9):727-731.
11. Samah OA, Ibrahim N, Alimon H, Karim MIA. 1993. Studies on fermentation products of cocoa beans. World Journal of Microbiology and Biotechnology, 9(3):381-382.
12. Saré AZ. 2001. Etude des Conditions d'Optimisation de la Couleur du Beurre de Cacao. Mémoire de fin d'études pour l'obtention du diplôme d'agronomie approfondie (DAA) de l'Ecole Supérieure d'Agronomie (ESA), Institut National Polytechnique Félix Houphouët Boigny de Yamoussoukro, Côte d'Ivoire.
13. SAS Institute , Inc. 1985. SAS User's Guide Statistic, 5th ed. SAS Inst., Cary, NC.
14. Selamat J. 1998. Acidic characteristics of cured and roasted cocoa beans from different countries of origin and flavor of the resulting chocolate. Dissertation Abstracts International B, 49(4)964: order number DA88O7851, page 188.
15. Schwan RF, Rose AH, Board RG. 1995. Yeasts, lactic and acetic bacteria, aerobic spore forming bacteria and fungi microbial fermentation of cocoa beans, with emphasis on enzymatic degradation of the pulp. Symposium series. Society for Applied Biotechnology, No. 24, 96S-107S.

Chapter 11

Teractiv Cacao as a New Fertilizer based Reactive Phosphate Rock for Cocoa Productivity in Côte d'Ivoire: A Participatory Approach to Update Fertilization Recommendation

Louis Koko

CNRA, Programme cacao, BP 808 Divo, Côte d'Ivoire

ABSTRACT

Cocoa (*Theobroma cacao L.)* is a strategic product for Côte d'Ivoire. This country is the world leader producer with aver 40 % of world volume production. This production is obtained on 6 % of the country's area by the work of 5 million farmers and the members of their family. Cocoa contributes to 30% of export earnings and 15% of the Gross Domestic Product (GDP). These economic parameters show the importance of cocoa for Côte d'Ivoire. Despite some remarkable performance, in cocoa growing systems, there are many constraints for cocoa productivity and sustainability. Particularly, in acid strongly P-sorbing soils of Côte d'Ivoire, P is among the nutrients that most limits cocoa production. For the success of cocoa planting, application of P to soils is one solution to ensure cacao growth and productivity. However, the current recommendations of fertilization based on the use of the fertilizer 0 N. 23 P_2O_5.19 K_2O^+ 10 CaO^+ 5MgO is not adapted in a context of deteriorating soil fertility. Improving the productivity of cocoa in Côte d'Ivoire therefore requires to update the fertilizer recommendations. In this context, the National Agronomic Research Center (CNRA, Côte d'Ivoire) conducts research-development in collaboration with the company Mars Inc. and Office Chérifien of Phosphates (OCP, Morocco), in order to test and validate a new fertilizer called Teractiv Cacao (0 % N. 15% P_2O_5. 14 % K_2O^+ 28% CaO^+ 5,5%S + 2,5 % MgO^+ 0,9 % Zn^+ 0,24 % B_2O_3). This new fertilizer includes natural tender phosphate (Reactive Phosphate Rock, RPR) in its composition, which makes it reduces the cost of purchase for cocoa producers and optimizes phosphate fertility of soils in cocoa growing systems. In its implementation, the research-development project adopted a participatory approach. Trials are conducted in three main cocoa regions in Côte d'Ivoire, including CNRA research stations Divo (Central - West) and Abengourou (East), as well as cocoa farmers areas in Soubré (South West). Ultimately, this project will contribute to the updating of fertilizer forms and doses in order to improve cocoa productivity in Côte d'Ivoire.

INTRODUCTION

Côte d'Ivoire is the world's largest producer of cocoa with an estimated annual production of more than 1.3 million tones, which represents 40% of world supply. However, the system of cocoa is extensive [1, 2, 3, 4]. More than 2,176,000 ha of agricultural area are devoted to cocoa [4]. In this system, about one million producers use non selected plant material, described and agricultural techniques are inefficient. Therefore, the orchard is very low productive. Average yields are between 260 and 600 $kg.ha^{-1}$ [3, 4, 5]. Today, the context of cocoa in Côte d'Ivoire is facing the loss of forest, traditionally used by producers as the best previous crop [6[. Plantations are mostly in a senescence phase after decades of exploitation ([5, 7]. Face degradation plantations, farmers began to plant cacao on previous non-forest (old cocoa plantations or coffee, natural fallow), but they face considerable difficulties. Very often, these constraints lead failures replanting [8[. A major constraint for the cocoa replanting is soil fertility declining [9]. In these soils, there is a decline in the level of key nutrients such as N, P, Ca, Mg, Fe and Zn after a long land use [10]. This came mainly due to the non-existence or inadequacy of the application of fertilizers and nutrient losses by erosion and leaching [10[. Soils also become more acidic, which causes interference to the availability and the uptake of P by cocoa [11, 12, 13, 14].The low availability of P in soils under cocoa is due to low levels of P naturally observed in soils, but also is due to the lack of application of phosphate fertilizers [13, 15[.

The practice of liming in order to raise the soil pH and precipitate exchangeable aluminum has been reported as an effective solution to reduce the effect of aluminum toxicity on P availability in acid soils [16]. But this practice requires large amounts (2 to 10 t.ha-1) of $CaCO_3$, which is not always accessible to small producers. However, in these soils, the problem of adsorption of P and the ineffectiveness of phosphate fertilizers has been demonstrated. In fact, 85-90% of inorganic phosphates fertilizers are applied in acid soils become unavailable to the plants during the year due to the adsorption and precipitation of P with Fe, Al and Ca in the soil [17]. Whereas the use of fertilizers is essential to intensify cocoa production, it is necessary to find alternatives that allow efficiency in the use of phosphate fertilizers. In Côte d'Ivoire, agricultural research has set a goal of improving cocoa production system in developing techniques cocoa both sustainable and adapted to the context of production [18,19]. The importance of fertilization in cacao cultivation is undeniable, and many studies have been conducted to establish fertilization techniques that allow improved planting material to express its full production potential (2-3 t.ha-1) [20]. However, studies have shown that low fertilizer recommendations are adopted by producers [21, 4]. Several reasons for this low level of adoption of fertilizer producers: insufficient knowledge or a lack of dissemination of fertilizer formulas developed, inefficient old formulas of fertilizers and excessive cost of fertilizer. However, improving the productivity of cocoa in Côte d'Ivoire requires updating fertilizer recommendations. In this context, the National Agronomic Research Center (CNRA, Côte d'Ivoire) is conducting an initiative on research and development in collaboration with the company Mars Inc. and the Office Chérifien des Phosphates (OCP, Morocco), to test and validate a new fertilizer called Teractiv Cocoa (N 0 %. 15 % P_2O_5. K_2O 14 % + 28 % CaO + MgO + 5.5 % + 2.5 % S 0.9 % Zn + 0.24 % B_2O_3).

Fertilizer includes natural tender phosphate (Reactive Phosphate Rock, RPR) in its composition, which makes it reduces the cost of purchase for cocoa producers and optimizes phosphate fertility of soils in cocoa growing systems. This paper aims to summarize the Ivorian experience of cocoa fertilization.

Three points are presented: The first provides a brief summary of the environmental requirements of cocoa and the role of fertilizer in cocoa production. The second point is devoted to the main scientific and technical achievements of the mineral fertilization of cocoa in Côte d'Ivoire. This point also raises the question of updating old fertilizer formulas. Finally, the third part of this paper presents one participatory strategy to update cocoa fertilizer formulas through the project Teractiv cocoa.

Environmental Requirements of Cocoa and Importance of Fertilization

Cocoa is a plant of humid tropics. Culture is particularly adapted to latitudes near the equator, 10 ° N to 10 ° S, and an annual rainfall of 1 500 to 2 000 mm, with less than three consecutive months with rainfall amounts of less than 100 mm. Cocoa supports average maximum temperatures between 30 ° C and 32 ° C and average minimum of 18 ° C to 21 ° C, with an absolute minimum of 10 ° C. The cocoa should be avoided in very windy areas, where cacao can suffer heavy defoliation in the absence of windbreaks [22]. Cocoa has very pronounced soil requirements [1, 9]. Indeed, it requires deep soil (minimum 1.5 m), well-drained, preferably sandy loam. The pH close to neutral (between 6 and 7.5), a topsoil rich in organic matter (30 g kg -1 minimum) and very rich in minerals are most favorable to cocoa. In Ivory Coast, the soils of the West and Southwest, which does not meet all these criteria, are sometimes the cause of the premature degradation of the orchard [9[. In addition, many other factors such as the type of plant material, repeated droughts or floods, insect attacks and cankers on trunks, cause exhaustion or death of productive cocoa. Finally, the sustainable production of cocoa for 25 to 30 years largely depends on soil fertility, hence the importance of mineral fertilization [23]. This follows three principles fertilization: Fertilization must first correct the deficits and imbalances in soil nutrients to allow optimum growth of cocoa, even on poor soils. Second, the role of fertilization increased nutrient levels to obtain high yields (more than 2 $t.ha^{-1}$). Finally, fertilization must be maintained to replace the nutrients exported by crops [1, 23].

MINERAL FERTILIZATION COCOA: FROM THE PAST TO THE PRESENT STAGE

Cocoa Fertilization by "Diagnosis Soil"

Mineral fertilization cocoa was first studied with the aim of reasoning fertilizer applications based on soil analyzes. Such applications should help restore some chemical equilibrium in soil to ensure a good cocoa production [24, 25, 26]. The dose calculation and determination of the fertilizer formulas are performed using computer software referred to as "soil Diagnosis", whereby, from a database of soil analysis, calculate chemical equilibrium following:

- Content of phosphorus (Olsen-Dabin method) equal to or greater than 60 $cmol.kg^{-1}$ P_2O_5;
- Total nitrogen balance - total phosphorus (N/P) of less than 2;
- A balance established between the exchangeable bases (BE) and total nitrogen (N), according to the formula $[BE + 6.15] /N = 8.9$;
- Balancing cations, expressed in percentage of the sum, almost 8% of K, 68% Ca and 24% Mg;
- A degree of base saturation of 60% minimum.

Fertilization trials from the "diagnostic soil" have been installed in different cocoa growing regions, in CNRA research stations, including Divo (Central West), Abengourou (East), Zagne (West) and Soubré (Southwest). Test plots were divided into two blocks: control and the other receiving fertilizer formulas and fertilizer doses calculated after "diagnosis soil."

Cocoa Fertilization by Simplified Formula

In 1985, looking for a simplified formula has been testing the research station, in collaboration with firms producing and marketing of fertilizers. The aim was to provide a formula of fertilizer to producers with agricultural efficiency is identical to that obtained with the method "Diagnosis Soil." After experimentations on research stations, gains cocoa production by at least 40% were obtained with a wide range of fertilizer formulas tested. Research has validated and recommended a simplified formula fertilizer called "cocoa fertilizer" 0N. 23 P_2O_5.19 K_2O + 10 CaO + 5MgO.

UPDATING FERTILIZER FORMULAS: INNOVATION WITH TERACTIV

However, due to the acidification of soils in cocoa growing systems (pH between 4.5 and 5.5), the development of efficient fertilization involves the search for new sources of P for acid soils. Indeed, sorption-desorption experiments conducted recently in Côte d'Ivoire showed that there is a variability in the intensity of sorption, depending on fertilization practices [27, 28, 29, 30[. Despite application of phosphate fertilizer in the form of triple super phosphate (P_2O_5 form as TSP) phosphate nutrition in cocoa growing systems is significantly limited by the high sorption capacity of P in soils (Fig 1).These studies showed that the sorption phase, characteristic of a solid phase with two types of reaction sites involving phosphate ions : a stability phase of the sorption curve for P inputs below 40 mg.kg-1 of P and a growth phase of the sorption above 40 $mg.kg^{-1}$ of P inputs [27, 28, 29, 30]. According to some authors [27, 28, 29, 30], soils Ferralsol (FAO soils classification) under cocoa have levels of exchangeable aluminum

Al_3^+ beyond 1cmol$^+$ kg^{-1} soil. They are not saturated after clearing and moderately saturated in cultivated soils [28, 29]. This strong presence of aluminum in solution and on the adsorbent complex is the origin of the high acidity of the soil and their high adsorption capacity of phosphate ions. These soils high variable load are among the most filled in total P (average grades of 300 to 1200 mg.kg^{-1} of P in soil), but also the poorest among the available P [30]. Increasing the concentration of phosphate ions in the solution of P inputs is not economically feasible because of the strong P sorption capacity in soils under cocoa [28, 29, 30]. To increase the bioavailability of P in these soils, reasoned farming practices should aim to reduce the holding capacity of P in the soils.

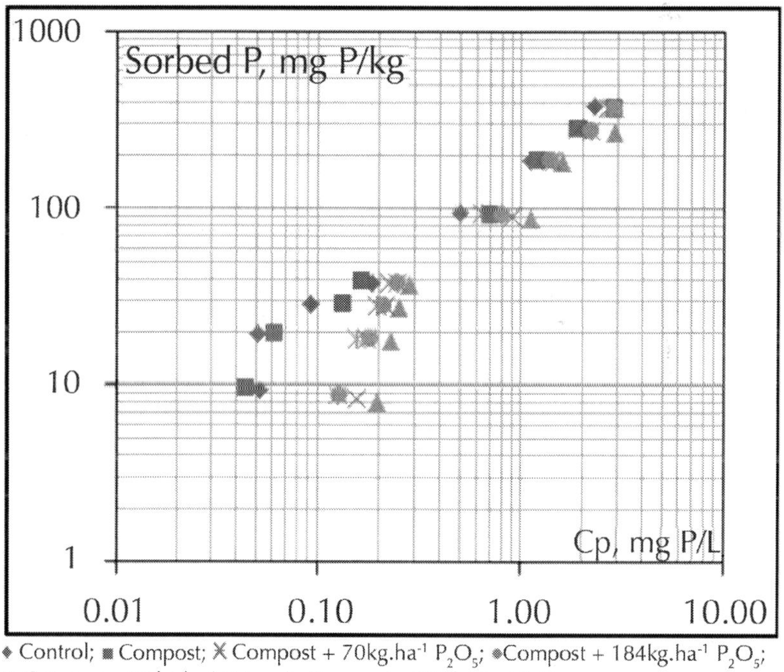

Figure 1: P sorption capacity in cocoa growing soils according to the integrated fertilization practices in Cote d'Ivoire [30].

In this context, the fertilizer formula Teractiv cocoa 0N -15 P_2O_5 -14 K_2O + 28 CaO + 5.5 S + 2.5MgO$^+$ 0.9Zn + 0.24 B_2O_3 was developed. This fertilizer provides some innovations:

- The raw phosphate is provided by a Reactive Phosphate Rock (RPR). This RPR is less costly and provides three times more Ca for two P_2O_5. In the old cocoa fertilizer formula 0 N. 23 P_2O_5.19 K_2O^+ 10 CaO^+ 5MgO, phosphorus is supplied in the form of triple superphosphate (TSP). The TSP more expensive provides for one Ca for two P_2O_5;
- Teractiv cocoa fertilizer improves phosphorus availability through the action of complexing calcium (provided in the form of calcium sulphate) against the aluminum and iron, thus releasing the binding sites of soluble phosphorus in soils;
- Fertilizer Teractiv cocoa enriched zinc and boron; these trace elements are necessary for strengthening the metabolism and the cacao fruiting.

These innovations could lead to improved mineral nutrition and productivity of cocoa. It is therefore necessary that research evaluates the effect of cocoa

Teractiv from demonstration plots on research stations and in farmers. It is in this context that the fertilizer Teractiv cocoa was proposed by OCP to study its effectiveness. This study, which began in 2012, will be conducted in three main cocoa growing regions of Côte d'Ivoire, including Divo (Central - West) and Abengourou (East), where the test will be conducted on research stations of the CNRA, and in the region of Soubré (Southwest) where the test will be implemented from producers through a participatory approach (Fig. 2). The project Teractiv cocoa will last 3 years and will aim to assess the effectiveness of the fertilizer Teractiv on cocoa productivity and improving soil fertility. Strategy implementation of the project Teractiv is participatory. At Soubré, cocoa farmers will be closely involved in the conduct of experiments so that they take ownership directly results to be obtained. This project will update the formulas and doses of fertilizer in cocoa production. Thus, the effect of fertilizers Teractiv on the cocoa yield will be known, the chemical fertility of soils under cocoa after fertilizer application Teractiv cocoa will be evaluated and the economic profitability of fertilizer application will Teractiv cocoa determined.

CONCLUSION

The continued leadership of the Côte d'Ivoire in cocoa production concerns all national actors in the cocoa. The aging of the orchard, the decline in soil fertility and the emergence of new diseases are the constraints for a drop in productivity. Today, producers are convinced of the importance of fertilizers to improve their production. This suggests an increase in the rate of fertilizer utilization in cocoa systems. A main challenge is therefore to improve the productivity of orchards mineral fertilization. To achieve this goal, the agricultural research actors are mobilizing. The National Agronomic Research Center (CNRA, Côte d'Ivoire) conducts research-development in collaboration with the company Mars Inc. and Office Chérifien of Phosphates (OCP, Morocco), in order to test and validate a new fertilizer called Teractiv Cacao (0 % N . 15 % P_2O_5. 14 % K_2O^+ 28% CaO^+ 5,5 % S + 2,5% MgO^+ 0,9% Zn^+ 0,24% B_2O_3). This new fertilizer includes natural tender phosphate (Reactive Phosphate Rock, RPR) in its composition, which makes it reduces the cost of purchase for cocoa producers and optimizes phosphate fertility of soils in cocoa growing systems. In its implementation, the research-development project adopted a participatory approach. Trials are conducted in three main cocoa regions in Côte d'Ivoire, including CNRA research stations Divo (Central - West) and Abengourou (East), as well as cocoa farmers areas in Soubré (South West). Ultimately, this project will contribute to the updating of fertilizer forms and doses in order to improve cocoa productivity in Côte d'Ivoire.

REFERENCES

1. Hanak-Freud E.H., Petithuguenin P., Richard J., 2000. Les champs de cacao. Un défi de compétitivité Afrique Asie. Karthala, Paris, 200 p.
2. Deheuvels O., Assiri A.A., Petithuguenin P., Kébé B.I., Flori A., 2005. "Production cacaoyère en Côte d›Ivoire : état actuel du verger et pratiques paysannes", Proceedings of the 14th International Cocoa Research Conference: towards a sustainable cocoa economy - what strategies to this end?Accra, Ghana, 2003, Cocoa Producers' Alliance, Lagos, [CD-ROM: 1157-1175.

3. Assiri A.A., 2007. Identification des pratiques paysannes dans la conduite des vergers de cacaoyers en Côte d'Ivoire. Mémoire de DEA, Université de Cocody, Abidjan, Côte d'Ivoire, 61 p.
4. Assiri A.A., Yoro G.R., Deheuvels O., Kébé B.I., Kéli Z.J., Adiko A. et Assa A., 2009. Les caractéristiques agronomiques des vergers de cacaoyer (Theobroma cacao L.) en Côte d'Ivoire. Journal of Animal & Plant Sciences, 2009, 2 (1) : 55-66.
5. Aguilar P., Paulin D., Keho Y., Nkamleu G., Raillard A., Deheuvels O. et Petithuguenin P., 2005. "L'évolution des vergers de cacaoyers en Côte d›Ivoire entre 1995 et 2002", Proceedings of the 14th International Cocoa Research Conference: towards a sustainable cocoa economy - whatstrategies to this end? Accra, Ghana, 2003, Cocoa Producers› Alliance, Lagos, [CD-ROM[: 1167-1175.
6. Gockowski J., Sonwa D.J., 2010. Cocoa Intensification Scenarios and Their Predicted Impact on CO_2 Emissions, Biodiversity Conservation, and Rural Livelihoods in the Guinea Rain Forest of West Africa. Environ Manag.
7. Kéli Z.J, Assiri A.A., 2001. "Essai de bilan et perspectives des systèmes de cultures associées à base de plantes pérennes en Côte d'Ivoire", Proc. Conférence sur l'Avenir des Cultures Pérennes : Investissement et durabilité en zones tropicales humides, 5-9 Novembre 2001, Yamoussoukro (Côted'Ivoire) – CD Rom.
8. Deheuvels O., 2007. Modélisation économique des exploitations agricoles : modélisation, simulation et aide à la décision avec le logiciel Olympe. In: Penot Eric (ed.), Deheuvels Olivier (Ed.). Paris : L›Harmattan, p. 49-61.
9. Koko L.K., Kassin K.E., Yoro G., NGoran K., Yao-Kouamé A., 2009. Corrélations entre le vieillissement précoce des cacaoyers et les caractéristiques morphopédologiques dans le Sud-Ouest de la Côte d'Ivoire. Journal of Applied Biosciences 24: 1508-1519.
10. Hartemink A.E., 2005. Nutrient stocks, nutrient cycling and soil changes in cocoa ecosystems : A review. Advances in Agronomy, vol. 8, p. 227-253.
11. Wessel M., 1971. Fertilizer requirements of cocoa (Theobroma cacao L) in South-Westem Nigeria. Communication 61, Department of Agricultural Research, Royal Tropical Institute, Wageningen, The Netherlands

12. Santana C.J.L, Cabala-Rosand P., 1984. Soil acidity and cacao response to lime application in South Bahia, Brazil. Proceeding of the 9th International Cocoa Research Conference, Lome, Togo, p. 199-203.

13. Cabala-Rosand P., Santana M.B.M., Santana C.J.L. de, 1989. Cacao. In : Detecting Mineral Nutrient Deficiencies in Tropical and Temperate Crops, Ed. DL Plucknett and HB Sprague, p. 409-425, Westview Tropical Agriculture Series, Westview Press, London.

14. Nakayama L.H.I., Pinto L.R.M., Santana, C.J.L., 1987. The effect of lime application on the cultivation of cacao (Theobroma cacao L), Proceedings of the 10th International Cocoa Research Conference, Santo Domingo, Dominican Republic, p. 259-264

15. Wilson K., 1999. Coffee, Cocoa and Tea, CAB1 Publishing, Oxon, UK.

16. Haynes R.J., 1984. Lime and phosphate in the soil–plant system. Adv Agron 37: 249–315.

17. Haynes R.J., Mokolobate M.S., 2001. Amelioration of Al toxicity and P defficiency in acid soils by additions of organic residues: A critical review of the phenomenon and the mechanisms involved. Nutr Cycl Agroecosyst 59: 47-63.

18. Kébé B., N'Guessan K.F., Tahi G.M., Assiri A.A., Koko L.K., 2009. Bien cultiver le cacaoyer en Côte d'Ivoire Fiche technique. Edition CNRA - CTA, p 4.

19. Kéli Z.J., Assiri A.A., Koffi NG., N'Goran J., Kébé I., 2005. Evolution de l'amélioration variétale du cacaoyer et des systèmes de production de la cacaoculture en Côte d'Ivoire. Sciences & Nature 2 (2): 209-218.

20. Koko L. K., Assiri A. A., Kassin K. E., Yoro G., 2010. Etat des lieux de la fertilisation en culture du cacaoyer : acquis vulgarisables et perspectives de recherche. Atelier de revue de la stratégie de fertilisation du caféier et du cacaoyer/FIRCA Projet « amelioration de la productivité des vergers et de laqualité des produits au champ » du 2QC. Yamoussoukro du 15 au 16 avril 2010. 12 p.

21. Koko K.L, Yoro R.G., N'Goran K., Assa A., 2008. Evaluation de la fertilité des sols sous cacaoyers dans le Sud-Ouest de la Côte d'Ivoire. Agronomie Africaine 10 (1): 81-95.

22. Mossu G., 1990. Le cacaoyer. Collection Le Technicien d'Agriculture Tropicale. Ed. Maisonneuve et Larose, Paris.159 p.
23. Petithuguenin P., 1998. Les conditions naturelles de production du cacao en Côte d'Ivoire, au Ghana et en Indonésie. Plantation, recherche, développement 5 (6): 393-405.
24. Jadin P., Snoeck J. 1985. La méthode du diagnostic sol pour calculer les besoins en engrais du cacaoyer. Café Cacao Thé 29 (4): 255-266.
25. Snoeck D., Abekoe M. K., Afrifa A. A., Appiah M. R. K., 2006. "The soil diagnostic method to compute fertilizer requirements in cocoa plantations", Proc. Int. Conf. on Soil Science, Accra, Ghana, 16-21 July 2006. 10 p.
26. Snoeck D., Afrifa A. A., Ofori-Frimpong K., Boateng E., Abekoe M. K., 2010. Mapping Fertilizer Recommendations for Cocoa Production in Ghana Using Soil Diagnostic and GIS Tools. West African Journal of Applied Ecology, vol. 17: 97-107.
27. Koko L.K., Ouattara T.V., Morel C., Snoeck D., Millner P., Than D., 2013. Experience of Côte d'Ivoire in using compost and fertilizer in combination to improve cacao establishment and productivity in Côte d'Ivoire. WCF/ACI sub-regional workshop on soil fertility management in cocoa cultivation. 26-28 February, Hotel Etoile du Sud, Grand Bassam, Côte d›Ivoire.
28. Koko L.K., Ouattara T.V., Morel C., Snoeck D., 2012. "Précocité de fructification et de production chez le cacaoyer : facteurs nutritionnels liés à la biodisponibilité et à la dynamique du transfert net des ions P dans les sols de Côte d'Ivoire", Proc. 17ème Conférence Internationale sur la Recherche cacaoyère, Youndé, Cameroun. Du 15 au 20 octobre 2012.
29. Koko L.K., Ouattara T.V., Morel C., Snoeck D., 2012. Integrated soil fertility management (ISFM) to improve cacao establishment and productivity in Côte d'Ivoire. World Cocoa Foundation (WCF) workshop for Alumni of the Norman E. Borlaug International Agricultural Science & Technology Fellowship Program – Global Cocoa Initiative Nadasah Royal Hotel, Accra September 3 – 8.

30. Koko L.K., Milner P., Thanh D., 2012. Amelioration of P deficiency in acid cocoa soils by additions of composts and different organic fertilizers. Final presentation of Cocoa Borlaug Fellowship at USDA-Foreign Agricultural Service Borlaug Office, Washington, DC (USA). July 20, 2012.

Chapter 12

Comparison of Antioxidant Activity and Flavanol Content of Cacao Beans Processed by Modern and Traditional Mesoamerican Methods

Elizabeth Chin[1], Kenneth B Miller[3], Mark J Payne[2], W Jeffery Hurst[2], and David A Stuart[2]

Corresponding author: W Jeffery Hurst

[1]Current address: Department of Food Science and Technology, University of California-Davis, Davis, CA 95616, USA

[2]The Hershey Center for Health and Nutrition, The Hershey Company, 1025 Reese Avenue, Hershey, PA 17033, USA

[3]1025 King Street, Palmyra, PA 17078, USA

ABSTRACT

The use of cacao, in its familiar food and beverage form, dates back about 3,800 years. By the time of the discovery of America by Europeans, the Aztecs and Mayans had developed methods for drying, roasting and grinding cacao beans for use in beverages and foods. Today, the same processes are used, but have been adapted to modern machinery. The current study was conducted to compare the traditional Mesoamerican processing and modern processing methods and their impact on the antioxidant activity and flavan-3-ol content of the resulting processed cocoa mass. Two cocoa bean types were used: unfermented cocoa beans commonly produced and consumed today by the native peoples of Mesoamerica, and fermented Ivory Coast cocoa beans representing the most common country of origin for cocoa in the world today. Both bean types were processed by roasting and grinding using traditional and modern processing methods. The traditional method consisted of roasting whole beans on an earthenware comal heated with wood charcoal, de-shelling the beans, then grinding the beans to a paste using a stone metate and mano. The modern method consisted of roasting cocoa beans with pilot-scale equipment that simulates large-scale chocolate manufacturing using whole bean drum roasters, deshelling the beans followed by mechanical grinding using a ball mill producing a chocolate paste. The antioxidant capacity and flavanol content of these chocolate pastes were then determined. The data show that the antioxidant capacity and flavanol content of unprocessed Mexican Lavado cocoa beans were significantly higher than that of the unprocessed, fermented Ivory Coast cocoa beans. In unfermented Mexican Lavado cocoa beans, the traditional and modern processing revealed no clear advantage with respect to the antioxidant and flavanol contents. However, for fermented Ivory Coast beans, there were some differences in the level of flavanols between traditionally processed and modern processed fermented beans. These differences, while significant for some, but not all of the flavanol measures, were attributed to final roast temperatures achieved and to the possible lack of uniformity of bean fermentation.

INTRODUCTION

Cacao originated and evolved in the tropical rainforests of the upper Amazon and Orinoco Rivers in South America [1]. When fresh fruit of cacao, also known as cocoa or cacao pods, are harvested, the hard woody pod is broken open and the cacao seeds, better known as cocoa beans, with their surrounding white, sweet pulp, are revealed. Fresh cacao beans at this stage taste bitter and contain little or no cacao flavor [2-4]. The archeological evidence dating back to about 1900 to 1400 BC, indicates that cacao juice expressed from the sweet pulp was the first product of the cacao seed to be used [5]. The archeological records suggest that the first consistent use of the brown, roasted cacao paste that was made into cocoa drinks, sauces and solid cocoa-containing foods dates back to about 1,000 BC [6]. The forms of cocoa that we are familiar with today have these brown cocoa solids —chocolate liquor and cocoa powder—which are the ingredients of chocolate candy, cakes, cookies and hot cocoa beverages.

While no one really knows how the processing of cocoa beans by fermentation, roasting and grinding actually developed, Edgar [7] speculates that Mayans routinely fermented cocoa beans for one to two days with their pulp to produce alcohol. He then speculates that if the Mayans separated the juice and dried the beans, the beans could be subsequently roasted on a comal found in most kitchens used to cook maize tortillas and stews (Figure 1) to produce cocoa aroma and beans with less bitterness. Indeed modern science has learned that fermentation lowers the bitterness of roasted cocoa beans as well as their content of flavan-3-ols [8, 9] while also creating more chocolate flavor when the beans are roasted [4]. Roasting also makes the inedible shells brittle and easy to remove. To make chocolate paste, Edgar speculates that roasted beans were ground using a kitchen grinding stone called a metate and a stone mano (Figure 2) which were also commonly available in households to make maize flour. We refer to this as the Traditional Process. These same processes of roasting, shell removal and milling have been mechanized and we refer to this as the Modern Process.

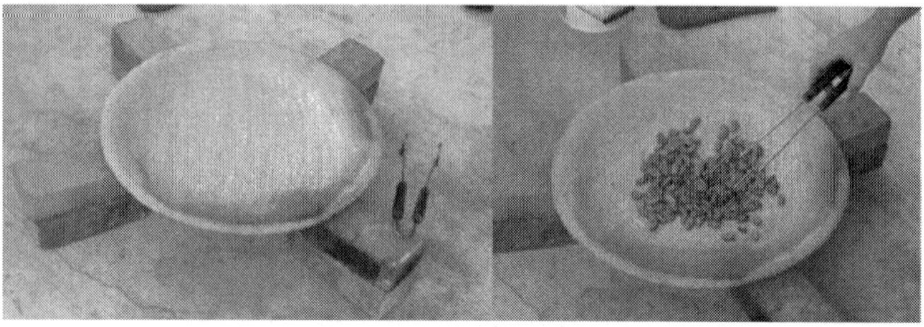

Figure 1: Traditional bean roasting. Comal (Left) with bricks to elevate it above charcoal heat from beneath. Bean roasting (Right) in which 250 gm of Lavado beans are being prepared.

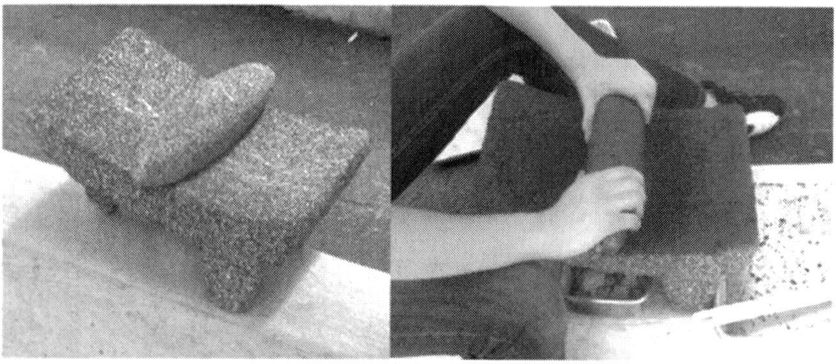

Figure 2: Traditional bean grinding. The metate base with the mano pestle on its surface (left). Bean grinding (right) on the metate. Note the charcoal heat being applied from beneath of the metate.

Cacao is known to contain very high levels of flavan-3-ols (flavanols) that occur both as monomers of epicatechin and catechin and as polymerized flavanols, or procyanidins [9-13]. The flavanols are commonly associated with cardiovascular health benefits which include lower rates of morbidity and mortality due to stroke and heart attack, improved vascular endothelial function, reduction in blood pressure and increases in HDL cholesterol [14, 15].

Studies have shown that flavanols are lost during transformation of cocoa beans into chocolate. Therefore, one can ask the compelling question regarding the level of these flavanols: Do traditional processing methods for roasting cacao beans have an advantage over the modern methods used today to produce roasted cocoa beans for making chocolate, perhaps because the process is more natural or not mechanized? The objective of this study is to compare the level of antioxidant capacity, flavanol monomers as well as oligomers and total flavanols in cacao beans processed by traditional Mesoamerican methods to the levels found in cocoa beans processed by modern methods. Here we have studied two types of cacao beans: unfermented cacao beans from southern Mexico, locally known as Lavado beans (lavar, in Spanish, meaning to wash) and conventionally fermented Ivory Coast beans. Beans of each type then were processed in a modern pilot plant to simulate industrial processing and were compared to beans prepared by the traditional Mesoamerican method which included comal roasting, hand winnowing and stone grinding on a metate.

MATERIAL AND METHODS

Cocoa Bean Samples

Two types of cocoa beans were obtained for this study: Mexican unfermented Lavado and Ivory Coast fermented. The Mexican Lavado beans were obtained from AMCO (Veracruz, Mexico) and were processed according to the local Lavado tradition in the Chiapas, Mexico region. The beans were removed from ripe pods, washed by hand with water to remove most of the pulp, then sun-dried. This results in unfermented, terracotta colored beans which are typical from this region of Mexico (Figure 3). Ivory Coast beans were obtained as standard fair average quality beans through The Hershey Company's purchasing department. These beans typically are heap-fermented for four to six days and then sun-dried. Ivorian beans have the typical brown appearance of fermented beans (Figure 3, bottom, right). The dried Lavado and the Ivorian beans had a moisture content of between 5.5% to 6.5% and fat content of 51% to 52%.

Figure 3: Unfermented and fermented cocoa beans. Mexican Lavado beans (Top, center) with their characteristic terracotta color. In the foreground are Ivory Coast fermented beans (Left; light brown) and Jamaica fermented beans (Right; dark brown).

Traditional Processing Method

Cocoa beans were roasted using methods that resembled traditional, ancient processing and typical of those still used today in Central America for local use. A 50-cm diameter earthenware comal was obtained from a Mexican market and was supported by several bricks. The comal was heated from beneath with charcoal (Figure 1). Once the center of the comal reached 250–300°C surface temperature as measured with a RayTek Mini-Temp Model MT6 laser thermometer (Fisher Scientific), 250 g of whole, dried cocoa beans were placed on the comal and heated. The beans were stirred occasionally with a spatula to avoid burning the outer shell. When beans developed a characteristic cacao aroma, they were heated for another three to four minutes, removed from the comal and cooled. Final external bean temperatures were 125–130°C as measured by a RayTek laser thermometer. The shells of roasted beans were fractured into small pieces using a stone mano and removed by hand or winnowed by sifting and blowing away the lighter shells from the heavier cocoa nibs. Shell-free nibs were ground using a stone metate and mano purchased from Melissa Guerra Cookware (Figure 2; bottom). The metate was kept slightly warm by charcoal underneath. The nibs were ground several times to produce a smooth, melted paste with a glossy appearance. Several batches were processed, combined and three aliquots were taken for analysis.

Modern Processing Method

Modern industrial roasting was accomplished on a pilot plant scale using 1362 g of whole, dried cocoa beans in a drum-style US Roaster. Burner temperature was set at 150°C and 2% water was added at the beginning of the roast. The beans were roasted to a final moisture content of less than 1% with a bean temperature of about 155°C. Shell-free nibs were obtained using a bench-top BLT cracker and winnower. Nibs were made into liquor with a bench-top Netsch Ball Mill and ground to a 25 μm particle size. Several batches were processed, combined, and three aliquots were taken for analysis.

Sample Preparation

Composite samples from each type of roasting were analyzed, to avoid measuring batch to batch variability but rather focusing on the impact of the different processing methods. In all cases, unroasted and roasted cocoa beans by both traditional and modern processing methods had their shells removed and were ground to a consistent particle size. Prior to analysis for the three flavanol tests described below, the ground mass was defatted by extracting three times with hexane [16]. Final results are reported on per gram of original cocoa bean material.

Antioxidant Capacity Measurement

Total antioxidant activity of unroasted and roasted cocoa bean samples was measured by the ORAC assay. ORAC is a widely used fluorescent method for assessing antioxidant capacity in food and biological samples. It is based on the inhibition of a peroxy radical induced oxidation initiated by the thermally based decomposition of 2, 2′-azobis- 2- methyl- propanimidamide, using fluorescein as a fluorescent probe and Trolox as a standard substrate. The results are expressed as μmol Trolox equivalent (TE) per gram [17,18].

Flavanol Monomer Determination

The flavanol monomers (±)-catechin and (±)-epicatechin were measured by the HPLC method of Nelson and Sharpless [19] standardized with (+)-catechin hydrate and (−)-epicatechin using the method recently validated using a five-laboratory ring test [20]. We define and report measures of (±)-epicatechin as epicatechin and (±)-catechin as catechin.

Flavanol Monomer to Decamer Determination

The flavanol oligomers with degree of polymerization (DP) ranging from DP 1 to 10 were measured by the HPLC method of Robbins et al.[21] with each sample extract standardized using individual purified oligomeric standards isolated and characterized as described by Hurst et al.[22]. These standards were obtained from Planta Analytica (Danbury, CT). The data are reported as Flavanols DP 1–10.

Total Flavanol Determination

Total flavanols were measured by the DMAC method which has been validated by a three laboratory ring-test described by Payne et al. [23]. This method is based on the specific reaction of p-dimethylaminocinnamaldehyde with the flavanol structure. The test was standardized using procyanidin B2 purchased from Planta Analytica (Danbury, CT). The method is specific to monomeric and polymeric flavan-3-ols as well as to related compounds such as epigallocatechin gallate and epicatechin-3-gallate. Monomers, i.e., epicatechin and catechin, react more strongly than higher MW oligomers with DMAC. Thus mg/g may not change, but the various proportions of monomers to DP2 and higher may be altered with proces sing. These results are reported as Total Flavanols.

Statistical Analysis

The data presented in this paper represent analysis of three representative samples for each sample type. Values for antioxidant activity and flavanol contents are expressed as averages and standard deviations. Due to the small sample populations, p-values were calculated on the basis of a significant difference. Therefore, at a 95% confidence level, p-values ≤0.05 represent a significant difference between two sets of data.

RESULTS AND DISCUSSION

Comparison of Unprocessed Lavado and Ivorian Beans

Table 1 provides the antioxidant and flavanol content for the unroasted Lavado and Ivory Coast cocoa bean material. The current results agree with previously published literature [9-13], namely that less fermented Lavado nibs have higher ORAC values and more flavanol content than fermented Ivory Coast nibs as measured by flavanol monomers, flavanol DP 1–10 and by total flavanols. As the literature has shown, ancient cultures would most likely have used less-fermented, Lavado-like cocoa beans. In fact, today there are areas ranging from southern Mexico to Venezuela where Lavado beans are preferred for local use. The reason is that beans washed free of the surrounding pulp are more easily dried during the rainy season and less subject to spoilage. However, today, chocolate manufacturers in the developed world rely on fermented beans like those from the Ivory Coast, because fermentation creates more chocolate flavor upon roasting and reduces the bitterness [3,4].

Table 1: Antioxidant capacity (ORAC) and flavanol content comparison for unroasted, unfermented Lavado beans and unroasted fermented Ivory Coast beans used in this study

Bean type	ORAC (μmol TE/g)	±−epicatechin (mg/g)	±−catechin (mg/g)	Flavanols DP 1–10 (mg/g)	Total flavanols (mg/g)
Washed, Unfermented Lavado	737±32	6.53±0.25	0.17±0.02	37.18±0.38	38.67±1.15
Fermented Ivory Coast	420±44	1.53±0.06	0.09±0.00	10.18±0.09	14.33±0.58

Three composite samples of each bean type were analyzed and values are expressed as average and standard deviation.

Since the focus of this study was a comparison of processing methods, the data presented in the following tables and graphs compare only the impact of traditional and modern bean processing on various antioxidant and flavanol measurements. A comparison will not be made to the unroasted starting material because the difference in moisture contents would create an unfair comparison.

Processing of Lavado (Unfermented) Beans

The data in Table 2 and Figure 4 shows a comparison of the antioxidant capacity and flavanol content of Lavado beans roasted by the traditional and modern methods. The major and perhaps surprising observation is that there were no significant differences in antioxidant capacity, epicatechin, flavanols DP 1 – 10 and total flavanols. The catechin content of the modern processed beans, however, was significantly higher than traditionally processed beans. The results in Table 2 and Figure 4 are perhaps surprising since it has been found that increased roast temperatures can decrease flavanol content when temperatures in excess of 70°C are reached [13]. Since the modern processed beans reached a higher terminal temperature of 150°C compared to the 125°C to 130°C exterior temperature of the traditionally processed beans, we might have expected to see lower antioxidant activities and flavanol values in the modern processed Lavado beans, but this was not observed.

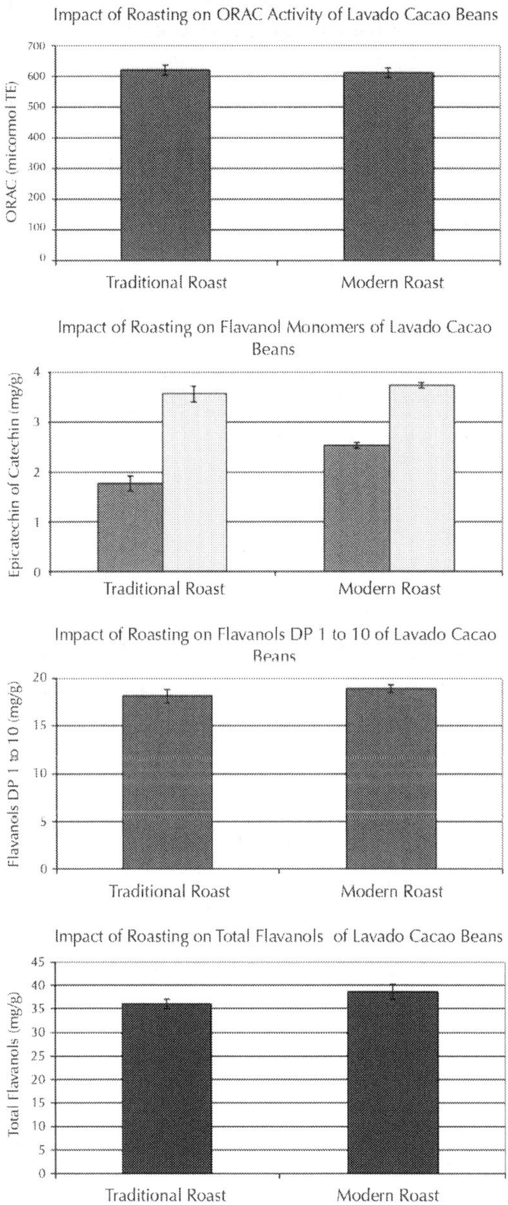

Figure 4: Comparison of unfermented (Lavado) cacao beans that have been left unprocessed, have been roasted and ground using traditional methods (Traditional Process) and have been processed using modern processing.

Top Graph. Analysis of antioxidant activity (ORAC); Second Graph. Analysis of epicatechin (yellow bars) and catechin (blue bars) content; Third Graph, Analysis of Flavanols DP 1–10 content and Bottom Graph, Analysis of Total Flavanol content.

Table 2: Antioxidant capacity and flavanol content comparison for unfermented Lavado cocoa beans roasted on a Mayan-style comal and ground on a metate (Traditional process) compared to the same beans roasted and ground using modern pilot plant equipment (Modern process)

Measure	**Lavado beans, traditional process**	**Lavado beans, modern process**	**p-value ***
ORAC (μmol TE/g)	620±17	613±15	0.6433
±−epicatechin (mg/g)	3.57±0.15	3.73±0.06	0.1518
±−catechin (mg/g)	1.77±0.15	2.53±0.06	0.0012
flavanols DP 1–10 (mg/g)	18.12±0.72	18.90±0.40	0.6143
total flavanols (mg/g)	36.00±1.00	38.67±1.53	0.0647

*The p-values calculated on basis of a 95% confidence of a significant difference at $p \leq 0.05$. Three composite samples were analyzed and values expressed as average and standard deviation.

Our results with unfermented beans are different from those of Cervelotti *et al.* in 2008 [24] who comparing the *in vitro* antioxidant activity and polyphenol content of finished chocolates made for standard, large scale manufactured products and for artisan manufacturing methods which used small batch roasting and stone, melangeur grinding. They concluded that artisan-made chocolate had higher antioxidant capacities when compared to mass produced chocolate as measured by Briggs-Rauscher reaction and by the Trolox Equivalent Antioxidant Capacity (TEAC) methods. They also found higher levels of polyphenols in the artisan chocolates. But their study made no attempt to measure the flavan-3-ol content of the chocolates, nor were they able to measure the antioxidant activities or the total polyphenol composition of the starting materials, as we have. Additionally, the Cervelotti study did not attempt to measure differences in their starting materials, nor differences in the processing methods used, since all of the chocolates produced would have used similar, but not identical, modern roasting and grinding techniques; the differences mainly being the scale of the manufacture. Our results with unfermented beans clearly show that there is a loss in antioxidant activity and flavanol content due to roasting but that there is no significant difference between the traditional process and the modern process in making chocolate mass.

Processing of Ivorian (fermented) Beans

In Table 3 and in Figure 5 are presented the effects of traditional and modern processing on the flavanol content of fermented Ivory Coast beans. Ivory Coast in 2009/10 represented the single largest origin of cocoa in the world producing 1.242 tones of cocoa or 34.1% of total world production [25]. The observations here are similar to the unfermented bean study. ORAC, epicatechin, and flavanols DP 1–10 were significantly higher in traditional processed beans by 20% ($p = 0.0021$), 26% ($p = 0.0075$), and 22% ($p = 0.0014$) respectively. Conversely, catechin and total flavanols trended higher in modern processed beans by 14% ($p = 0.060$) and 8% ($p = 0.089$) respectively, compared to traditionally processed beans. There are two possible explanations for these results. First, cocoa bean fermentations are done by very crude methods and the resulting fermented beans are highly heterogeneous with respect to degree of fermentation. This may impact the level of starting antioxidant capacity and flavanols in each bean. Therefore, one might expect more variability in the data from fermented beans compared to data from unfermented beans. Second, the higher terminal temperature in the modern processed beans partially, but does not fully, explain the selective loss of only antioxidant activity, epicatechin and flavanols DP 1–10 compared to the traditional process. But this does not explain why catechin and total flavanols are numerically higher (not significant at $p = 0.05$) when processed by the modern method. If roasting had been done to the same final temperature, it is likely that any differences might disappear [13]. Since these observations do not show a clear trend towards modern or traditional processing, it can be concluded that there is little difference in the *in vitro* antioxidant activity and flavanol contents of the more fermented Ivory Coast beans when processed in the traditional or modern way. This brings us to the same general conclusion that the traditional and the modern processes for roasting and grinding cocoa beans have about the same effect on the level of cocoa bean flavanols.

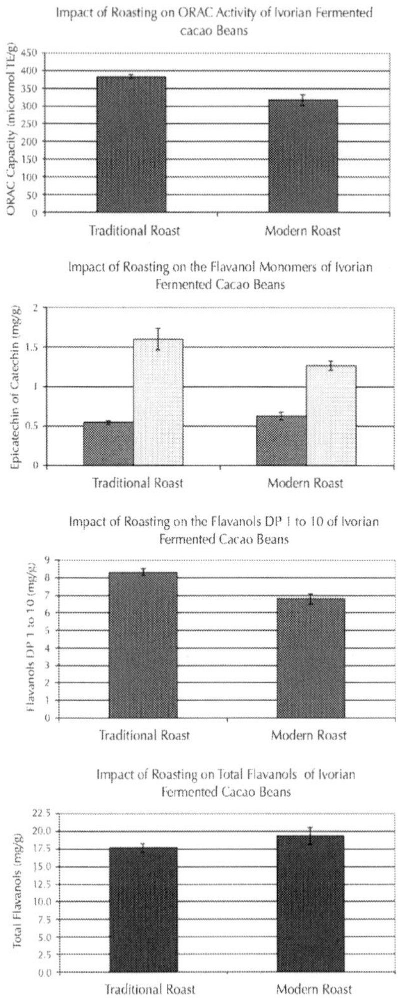

Figure 5: Comparison of fermented Ivory Coast (Ivorian) cacao beans that have been left unprocessed, have been roasted and ground using traditional methods (Traditional Process) and have been processed using modern processing.

Top Graph. Analysis of antioxidant activity (ORAC); Second Graph. Analysis of epicatechin (yellow bars) and catechin (blue bars) content; Third Graph, Analysis of Flavanols DP 1–10 content and Bottom Graph, Analysis of Total Flavanol content.

Table 3: Antioxidant capacity, and flavanol content comparison for Ivory Coast beans roasted on a Mayan-style comal and ground on a metate (Traditional process) compared to the same beans roasted and ground using modern pilot plant equipment (Modern process)

Measure	Ivorian beans, traditional process	Ivorian beans, modern process	p-value * (if different)
ORAC (μmol TE/g)	383 ± 6	317 ± 15	0.0021
± – epicatechin (mg/g)	1.60 ± 0.14	1.27 ± 0.06	0.0075
± – catechin (mg/g)	0.54 ± 0.03	0.63 ± 0.05	0.0596
flavanols DP 1–10 (mg/g)	8.30 ± 0.20	6.79 ± 0.30	0.0014
total flavanols (mg/g)	17.67 ± 0.58	19.33 ± 1.15	0.0890

*The p-values calculated on basis of a 95% confidence of a significant difference at $p \leq 0.05$. Three composite samples were analyzed and values expressed as average and standard deviation.

When the Lavado bean processing data (Table 2) are compared to Ivory Coast processing data (Table 3), the same observation can be made that the antioxidant capacity and flavanol contents of roasted, unfermented Lavado beans are significantly higher than in the roasted, more-fermented Ivory Coast beans (Table 1). This also is consistent with data published by Payne, et al. [13] and others [10-12] where the flavanol monomer and oligomer content of roasted Lavado beans has been found to be significantly higher than roasted beans from fermented sources like Ivory Coast and Papua New Guinea.

Implications for Traditional Processing and Modern Processing

People native to Mesoamerica and Northern South America have a tradition of consuming less-fermented or totally unfermented cocoa. This is due to the ease of processing and drying cocoa beans in humid climates, particularly during the rainy season, when the beans need to be quickly dried. Over the centuries these people have come to prefer the unfermented, Lavado beans. Regardless of the processing methods used, Lavado beans had the same level of flavanols and these levels were more than twice the level found in fermented beans after similar processing (compare Table 2 to Table 3). This may mean that people native to the Americas today and before the discovery of the New World by Columbus and Cortez, ate cocoa products made from cocoa mass (also called chocolate liquor) that contained about twice the level of flavanols found in today's cocoa mass.

After Europeans began importing cocoa beans, first for use in beverages, and then for solid chocolate, hot cocoa beverages and later milk chocolate, there was increased demand for partially or fully fermented cocoa beans. The producers of cocoa in the Americas were asked to produce cocoa beans that had been fermented to suit European tastes and were paid a premium for these beans [3,4]. As is shown in this study (Table 1), and by others [9-13], fermentation lowers the level of flavanols in the raw beans. The benefit to the European consumer was to have cocoa drinks or chocolate bars that were less bitter and had more cacao flavor. With the increased use of fermented beans and with the development of modern cocoa bean processing, there was a loss of flavanols from product, with the greatest loss of flavanols due to the preference for fermented beans. The results of the present study have shown that modern processing has little or no impact on flavanols (Table 3, Figure 5). Thus the health benefits of cocoa and chocolate flavanols and other beneficial bioactives need to be balanced with the flavor of the final product, for which fermentation of cocoa beans tends to increase cacao flavor impact and lessen bitterness. Despite the implementation of fermentation, dark chocolate and cocoa powder still have some of the highest concentrations of flavanols of any food in our diet [26]. In fact a recent dose response study using conventionally processed cocoa powder in ever increasing amounts has found that as little as 5 g is sufficient to elicit significant vasodilation of the brachial artery, as measured by Flow Mediated Dilation [27].

COMPETING INTERESTS

The authors declare that they have no competing interests.

AUTHORS' CONTRIBUTIONS

EC, KM and DS carried out cocoa roasting while MP performed and directed the analysis of samples. WH assisted in experiment design, sample processing and directed the project. All authors read and approved the final manuscript.

ACKNOWLEDGEMENTS

The authors would like to thank the Hershey Center for Health and Nutrition for providing the samples and facilities for this research. We thank The Hershey Company for supplying support to E. Chin as part of her undergraduate summer internship. Thanks go to John Scharffenberger for his expert advice and direction for the ancient, traditional processing of cacao beans. Also, a special thanks to W. J. Hurst for the acquisition of an authentic clay comal from Latin America.

REFERENCES

1. Powis : Cacao and the San Lorenzo Olmec. Proc Natl Acad Sci 2011, 108:8595-8600.
2. Coe SD, Coe MD: The True History of Chocolate. London, England: Thames and Hudson Ltd.; 1996.
3. Minifie BW: Chocolate, Cocoa and Confectionery. Westport, CT, USA: AVI Publishing Company; 1980.
4. Beckett ST: Industrial Chocolate Manufacture and Use. Blackie: Glasgow and London; 1988:388.
5. Henderson JS, Joyce RA, Hall GR, Hurst WJ, McGovern PE: Chemical and archaeological evidence for the earliest cacao beverages. Proc Natl Acad Sci USA 2007, 104:18937-18940.

6. Hurst WJ, Tarka SM Jr, Powis TG, Valdez F Jr, Hester TR: Cacao usage by the earliest Maya civilization. Nature 2002, 418:289-290.
7. Edgar B: The power of chocolate: tracking the chemical signature of cacao across Mesoamerica. Archaeology 2010., 63 Nov/Dec., 20–25
8. Thompson SS, Miller KB, Lopez AS: Chapter 39, Cocoa and Coffee. In Food Microbiology: Fundamentals and Frontiers, 3rd Ed. Washington DC: ASM Press; 2007.
9. Harborne JB, Baxter H, Moss GP: Phytochemical Dictionary: a handbook of bioactive compounds from plants. 2nd edition. London, United Kingdom: Taylor and Francis; 1999.
10. Kim H, Keeney PG: (−)-Epicatechin content in fermented and unfermented cocoa beans. J Food Sci 1984, 49:1090-1092.
11. Porter LJ, Ma Z, Chen BC: Cocoa procyanidins: Major flavonoids and identification of some minor metabolites. Phytochemistry 1991, 20:1657-1663.
12. Lea PJ, Ford GD: Characterization of the polyphenols of cocoa. Glasgow: BioFlavors Conference; 1990.
13. Payne MJ, Hurst WJ, Miller K, Rank C, Stuart DA: Impact of fermentation, drying, roasting, and Dutch processing on epicatechin and catechin content of cacao beans and cocoa ingredients. J Ag Food Chem 2010, 58:10518-10527.
14. Hooper L, Kay C, Abdelhamid A, Kroon PA, Cohn JS, Rimm EB, Cassidy A: Effects of chocolate, cocoa and flavan-3-ols on cardiovascular health: a systematic review and meta-analysis of randomized trials. Am J Clin Nutr 2012, 95:740-751. PubMed Abstract | Publisher Full Text OpenURL
15. Reid K, Fakler TR, Frank OR, Stocks NP: Effect of cocoa on Blood Pressure (Review). The Cochrane Collaboration. John Wiley Press; 2012:1-83.
16. Association of Analytical Chemists, International: Cacao mass (fat-free) of chocolate liquor. Official Methods of the AOAC International, 16th Ed. AOAC Official Method. Gaithersburg, MD: AOAC International; 1995. 931.05

17. Ou B, Hampsch-Woodill M, Prior R: Development and validation of an improved oxygen radical absorbance capacity assay using fluorescein as the fluorescent probe. J Agric Food Chem 2001, 49:4619-4626. PubMed Abstract

18. Huang D, Ou B, Hampsch-Woodill M, Flanagan J, Deemer EK: Development and validation of oxygen radical absorbance capacity assay for lipophilic antioxidants using randomly methylated-cyclodextrin as a solubility enhancer. J Agric Food Chem 2002, 50:1815-1821. PubMed Abstract

19. Nelson BC, Sharpless KE: Quantification of the predominant monomeric catechins in baking chocolate standard reference material by LC/APCI-MS. J Agric Food Chem 2003, 51:531-537. PubMed Abstract

20. Shumow L, Bodor A: An industry consensus study on an HPLC fluorescence method for the determination of (±)-catechin and (±)-epicatechin in cocoa and chocolate products. Chemistry Central Journal 2011, 5(39):MS ID: 7184481134944588.

21. Robbins RJ, Leonczak J, Johnson JC, Li J, Kwik-Uribe C, Prior RL, Gu L: Method performance and multi-laboratory assessment of a normal phase high pressure liquid chromatography-fluorescence detection method for the quantitation of flavanols and procyanidins in cocoa and chocolate containing samples. J Chromatorgr A 2009, 1216:4831-4840.

22. Hurst WJ, Stanley B, Glinski JA, Davey M, Payne MJ, Stuart DA: Characterization of primary standards for use in the HPLC analysis of the procyanidin content of cocoa and chocolate containing products. Molecules 2009, 14:4136-4146. PubMed Abstract

23. Payne MJ, Hurst WJ, Stuart DA, Ou B, Fan E, Ji H, Kou Y: The use of the DMAC assay for determination of total procyanidins in selected chocolate and confectionery products. J AOAC International 2010, 93:89-96.

24. Cervelatti R, Greco E, Costa S, Guerra M, Speroni E: A comparison of antioxidant properties between artisan-made and factory-produced chocolate. International Journal of Food Science & Technology 2008, 43:1866-1870.

25. ICCO statistics for World Cocoa Production. 2009. http://www.icco.org/statistics/production-and-grindings/production.html.

26. USDA Database for the Proanthocyanidin Content of Selected Foods. 2004. http://www.nal.usda.gov/fnic/foodcomp/Data/PA/PA.html.

27. Monahan KD, Feehan RP, Kunselman AR, Preston AG, Miller DL, Lott MEJ: Dose dependent increases in flow-mediated dilation following acute cocoa ingestion in healthy older adults.

Chapter 13

Environmental Impacts of Cocoa Production and Processing in Ghana: Life Cycle Assessment Approach

Augustine Ntiamoaha, George Afrane b,*

[a]Chemical Engineering Department, Kwame Nkrumah University of Science & Technology, Kumasi, Ghana

[b]Koforidua Polytechnic, Koforidua, Ghana

ABSTRACT

Ghana is the world's second largest producer of cocoa beans. In addition to exporting raw cocoa beans, the country also processes some of its beans into finished and semi-finished cocoa products for both the local and international markets. This paper is aimed at providing a comprehensive picture of the environmental impacts associated with cocoa production and processing in Ghana by applying the life cycle assessment (LCA) methodology. The analysis considered the entire system required to produce and process 1 kg of cocoa bean. It included the extraction of raw materials (e.g. fossil fuels, minerals), the production of farming inputs (e.g. fertilizers and pesticides) and all agricultural operations in the field (e.g. tillage, fertilizer and pesticides application, harvest, etc.). Transportation of beans to processing factory and industrial processing of the beans into cocoa butter, liquor, cake and powder were also included. The study was conducted in accordance with the international ISO procedural framework for performing LCA in the ISO 14040e14043 series. The overall environmental impacts resulting from cocoa production and processing activities and improvement options towards the sustainability of the system studied are presented and discussed.

INTRODUCTION

Cocoa, which is used mainly in the production of chocolate, is an important agricultural export commodity. Currently Ghana, producing about 700,000 tons of cocoa beans annually, is ranked second in the world, after her western neighbour Coˆte d'Ivoire. In terms of quality however, Ghana is recognized as the world leader in premium quality cocoa beans production.

Cocoa serves as the major source of revenue for the provision of socio-economic infrastructure in the country. In terms of employment, the industry employs about 60% of the national agricultural labour force in the country [1]. For these farmers, cocoa contributes about 70e100% of their annual household incomes [2].

Improving the livelihood of farmers is a crucial aspect of government's plans for reducing poverty in Ghana. It has therefore been the intention of government, which is committed to reaping the maximum benefit from the cocoa sector, to ensure that the country increases its cocoa production and also processes more of the beans into downstream products for both the local and export markets [3]. In pursuance of this goal, the government in year 2001 initiated a nationwide Cocoa Disease and Pest Control Project (CODAPEC), to help address the two major causes of decline in cocoa production: pests and diseases. Under this programme, cocoa farms across the country were sprayed with insecticides and fungicides at no cost to the farmers. This exercise has resulted in tremendous increases in cocoa production from 340,562 metric tons in the 2001/02 season to 496,846 metric tons in 2002/03 and 736,000 metric tons in the 2003/04 seasons, respectively [1,4].

The percentage of locally processed beans has also jumped from 20% to 35% with further re-capitalization and expansion programs underway to reach a target of 50% in the near future. The beans are normally processed into four semi-finished products, namely, cocoa liquor, cocoa butter, cocoa powder and cake; and a finished product, chocolate. However, along with the positive effects of the CODAPEC programme, some negative impacts on the environment have also been caused. For instance the use of pesticides on the farms can lead to the destruction of part of the soil flora and fauna through both physical and chemical deterioration [5].

This study was aimed at providing a comprehensive picture of the environmental impacts associated with the current cocoa production and processing techniques in Ghana, using the life cycle assessment (LCA) tool. LCA determines the potential environmental impacts, throughout a product's life cycle, from raw material acquisition through production, use, endof- life treatment, recycling and to final disposal (i.e. cradleto- grave) [6].

METHOD

The study was conducted in accordance with the ISO procedural framework for performing LCAs in the ISO 14040e 14043 series [6e9]. According to the ISO standards, LCA study has four main phases, namely, goal and scope definition, life cycle inventory analysis, life cycle impact assessment, and interpretation of results.

Goal and Scope of the Study

The goal and scope definition phase states why an LCA is being conducted and describes the system to be studied. The objectives of this particular study were to identify and quantify the potential environmental impacts associated with cocoa production and processing in Ghana, to identify the activities that are not performing sustainably (i.e. those with the largest environmental impacts) and then to suggest a number of improvement options or impact reduction strategies towards the sustainability of the system studied. The results of the study could be used by the Ghana Cocoa Board (COCOBOD) e the main government agency in charge of the cocoa industry e and other stakeholders like cocoa scientists, farmers and processors, in decision-making to improve the environmental performance of the industry.

The system studied was sub-divided into three main stages as follows: cocoa production (comprising pesticides and fertilizer production, cocoa cultivation, on-farm processing and temporary storage of cocoa beans); transportation (involving transport of beans to processing factory); and industrial processing of cocoa beans (comprising beans-cleaning, roasting, breaking and winnowing, and grinding into cocoa liquor, cocoa butter, cake and cocoa powder). Life cycle tracing of all packaging materials was excluded from the study due to lack of relevant data.

Functional Unit

LCA is a relative approach which is structured around a functional unit. This functional unit defines what is being studied [6]. For this study, the functional unit chosen was 1 kg of cocoa beans processed. All the inputs and outputs in the life cycle inventory (LCI) and impact scores produced in the life cycle impact assessment (LCIA) phase of this LCA study were expressed with reference to this functional unit.

Life Cycle Inventory (LCI) Analysis

The life cycle inventory (LCI) analysis segment is the second phase of LCA. It is an inventory of input/output data with regard to the system being studied. The inventory analysis consists of two major steps: data collection and data analysis.

Data Collection

Data for cocoa production were collected from government records on cocoa production in Ghana, and also by paying site visits to farms and institutions and interviewing cocoa farmers and researchers at the Cocoa Research Institute of Ghana (CRIG). Background data on production of fertilizers and pesticides, transportation and electricity generation were included using the Swiss eco-invent database and the GaBi 4 LCA database.

Emissions due to fertilizer and pesticide usage were quantified using estimation methods [10,11]. Inventory data for transportation was calculated based on average distance of 250 km traveled by diesel engine trucks in Ghana. The truck used for the analysis was assumed to be 22 tons total capacity/ 14.5 tons payload/long distance truck. A state-owned cocoa processing company, considered state of the art, was selected for detailed study and data gathering on cocoa processing. A summary of the inventory data collected is presented in Table 1.

Table 1: Summary of inputs and outputs data for the production and processing of 1 kg cocoa bean in Ghana, for the reference year 2004/2005

Inputs/outputs	Amount	Unit
Energy inputs		
Electricity (from national grid)	3.1716E-01	MJ
Diesel	5.3142E-02	kg
Petrol	8.9967E-03	kg
Materials inputs		
Water	5.1274E-00	kg
Fertilizer (N:P:K 0:22:18 þ 9CaO þ 7S þ 6MgO)	1.4590E-01	kg
Pesticides		
-Fungicides	7.4200E-03	kg
-Insecticides	8.0000E-04	kg
Land use	3.9218E-05	ha

Products/by-products		
Cocoa liquor	3.1948E-01	kg
Cocoa butter	2.3125E-01	kg
Cocoa cake	2.6875E-01	kg
Cocoa powder	7.5000E-02	kg
Cocoa shells	9.8000E-02	kg
Air emissions		
Dust (PM2,5 e PM10) [particles to air]	2.5000E-03	kg
Sulphur dioxide [inorganic emissions to air]	8.0000E-03	kg
Heavy metals to air	3.5745E-05	kg
Carbon dioxide [inorganic emissions to air]	2.3790E-01	kg
Carbon monoxide [inorganic emissions to air]	8.4100E-03	kg
Pesticides to air	8.1308E-04	kg
Water emissions		
Biological oxygen demand (BOD)	5.0437E-12	kg
Chemical oxygen demand (COD)	9.8212E-12	kg
Nitrates	3.7500E-15	kg
Oil and grease	1.0000E-14	kg
Phosphates	4.4204E-14	kg
Total dissolved solids	5.1525E-12	kg
Total suspended solids	4.1287E-12	kg
Heavy metals to freshwater	7.4761E-04	kg
Pesticides to freshwater	3.6880E-03	kg
Soil emissions		
Pesticides to soil	9.4477E-04	kg
Heavy metals to agricultural soil	4.1870E-05	kg

Data Analysis

The data collected were converted to values that relate to the functional unit. The adjusted data were entered into the GaBi 4 LCA software and modeled into environmental inputs and outputs and then aggregated to result in an inventory table.

Life Cycle Impact Assessment

This step calculates the likely human and ecological effects of material consumption and environmental releases identified during the inventory analysis. Classification and characterization following the ISO 14042 [8] guidelines were applied to the inventory data in order to assess their potential impacts on the environment. According to these guidelines, four optional elements, namely, normalization, valuation, grouping and data quality analysis may be included in the impact assessment.

However, in this study the mandatory elements (classification and characterization) were believed to be sufficient to achieve the stated goals of the study and therefore only these were considered. In the classification step, the inventory data were grouped into the environmental impacts of global warming, atmospheric acidification, eutrophication, photochemical ozone creation, freshwater aquatic eco-toxicity, terrestrial eco-toxicity, human toxicity, and ozone layer depletion potentials and depletion of abiotic resources. In the characterization step the inventory data that have been classified under each environmental impact category were quantified in terms of a common unit for that category. This was done by multiplying the mass value of the relevant emission by its corresponding characterization (or equivalency) factor (provided with the GaBi 4 LCA software) to give an indicator result for that inventory item. All indicator results under each impact category were summed to result in an overall impact score (or characterization results) for the impact category. The impact assessment was done by using the CML 2001 method developed by the Centre for Environmental Science, University of Leiden.

RESULTS AND DISCUSSION

The characterization results for the production and processing of 1 kg cocoa in Ghana, based on the CML 2001 method, is presented in Table 2 and illustrated graphically in Fig. 1. As shown in Fig. 1, freshwater aquatic eco-toxicity, human toxicity and global warming potentials are the most significant environmental impacts associated with the system studied. Fig. 2 illustrates the relative contributions of each life cycle stage studied to the overall environmental impacts.

Table 2: Characterization results in absolute values for the production and processing of 1 kg cocoa in Ghana, based on the CML 2001 method

Environmental impact category	Total impact score	Unit
Abiotic Depletion Potential (ADP)	1.6184E-03	kg Sb-equiv.
Acidification Potential (AP)	8.4237E-03	kg SO_2-equiv.
Eutrophication Potential (EP)	1.0476E-03	kg PO_4^{3-} -equiv.
Freshwater Aquatic Eco-toxicity Potential (FAETP)	5.8496E-00	kg DCB-equiv.
Global Warming Potential (GWP 100 years)	3.2286E-01	kg CO_2-equiv.
Human Toxicity Potential (HTP)	5.1144E-00	kg DCB-equiv.
Ozone Layer Depletion Potential (ODP)	5.7284E-09	kg R11-equiv.
Photochemical Ozone Creation Potential (POCP)	8.0877E-04	kg C_2H_4-equiv.
Terrestric Eco-toxicity Potential (TETP)	7.1222E-03	kg DCB-equiv.

DCB: 1,4-dichlorobenzene, R11: trichlorofluoromethane.

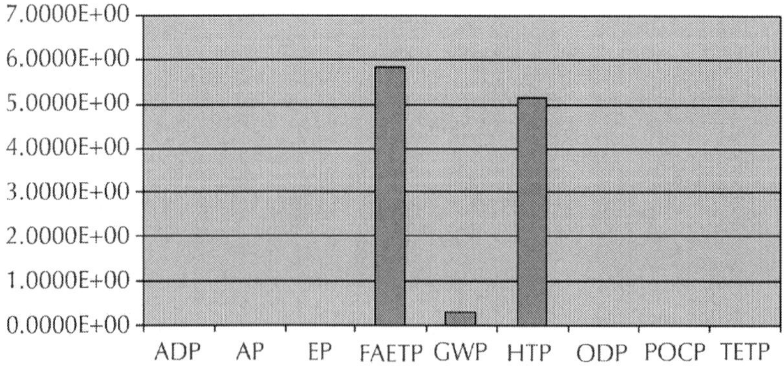

Figure 1: Characterized results for the production and processing of 1 kg cocoa in Ghana.

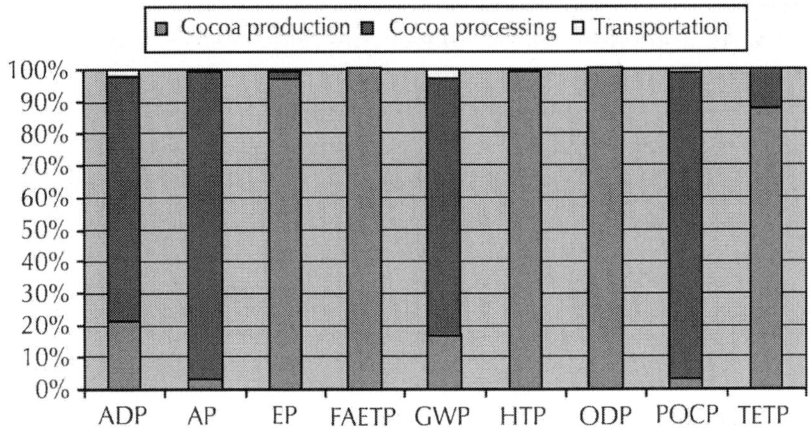

Figure 2: Contributions by different stages of production to overall impact scores.

Cocoa Production Stage

Fig. 2 shows that, cocoa production makes the largest contribution to the environmental impacts of eutrophication, ozone layer depletion, freshwater aquatic eco-toxicity, human toxicity, and terrestrial eco-toxicity, with average contributions greater than 96%. The analysis revealed that production and use of fertilizers and pesticides were a major cause of the environmental burdens in the cocoa production stage. Ozone layer depletion is caused by the emission of halogens and CFCs during the production of pesticides. Eutrophication is mainly caused by leakage of nutrients during cultivation and emission of phosphates from the production of phosphorus fertilizers.

For all the three toxicity categories the main contributors are heavy metals content in phosphorus fertilizers and leakage of pesticides. Agricultural machinery and implements are not used in cocoa production in Ghana because of the small farm holdings, and production largely depends on rainfall. Drying of the beans is by "sun drying". Therefore improvement measures should focus on reducing fertilizers and pesticides usage.

In addition to these impacts derived quantitatively by the LCA tool, there are other known important impacts that result from cocoa production. The cocoa production stage generates a large amount of solid waste in the form of pod husks. The pod husk constitutes about 67% of the fresh pod weight [12]. At present, pod husks are largely a waste product of the Ghanaian cocoa industry, and present a serious disposal problem. The direct impacts of these could not be assessed by the LCA tool. However, according to Figueira et al. [13] they became a significant source of disease when used as mulch inside the plantation. Cocoa seeds are surrounded by an aromatic pulp which arises from the seed teguments (technically an aril). During on-farm processing of cocoa beans, the pulp is removed by fermentation and is hydrolyzed by microorganisms.

Hydrolyzed pulp is known in the industry as "sweatings." Approximately 40 l of pulp can be obtained from 800 kg of wet beans [13]. The sweatings are also allowed to drain off as liquid waste in the farms. The sweatings and wastes from cocoa pod husks have already found uses and studies are being undertaken in Ghana to determine their viability as commercial products [12]. The commercialization of these new products could be an important economic incentive to cocoa production in the country.

Another important environmental issue of concern is loss of biodiversity. Large quantities of healthy and genetically diverse native flora and fauna are an indication of a balanced ecosystem [14]. The biodiversity loss of clearing original vegetation such as forest and the loss through the use of pesticides must be estimated. This indicator is now under development for the LCA methodology [15].

Cocoa Processing Stage

Industrial processing of cocoa beans was found to be the most predominant stage in the environmental impacts of photochemical ozone creation potential (95.84%), global warming potential (80.89%), atmospheric acidification potential (96.47%) and abiotic depletion potential (76.35%). The production and consumption of fossil fuels in boilers and roasters were identified as the main cause of environmental impacts in the cocoa processing stage. Therefore efforts must be focused on improving the efficiency of energy use in these energy-intensive equipments. It also proposed to substitute the use of diesel with natural gas since it is known to be relatively more environmentally friendly.

The main by-product from the processing industry, cocoa shell, is no longer considered a solid waste, as it is being processed and packed for sale as animal feed by the Cocoa Processing Company.

Transportation Stage

The transportation stage is relatively the most environmentally friendly of the sub-systems studied, since its contributions to the impact categories considered were not very significant.

IMPROVEMENT ANALYSIS

Proposed improvement options were modeled and their results compared with the baseline case to find out whether these modifications will reduce the environmental impact potentials. The baseline data depends on the present techniques and practices of cocoa production and processing in Ghana. The proposed improvement options tested are summarized in Table 3.

Table 3: Proposed improvement options

Life cycle stage	Base case	Improvement option suggested
Cocoa production	Mineral fertilizer	Organic fertilizer
Cocoa production	Mineral fertilizer	50% Reduction
Cocoa production	Pesticides	50% Reduction
Cocoa processing	Diesel oil	Natural gas

Each improvement option was tested independently of all others so that the extent of its effects on the reference case could be assessed alone. Figs. 3e6 compare the environmental scores of these modifications with the baseline.

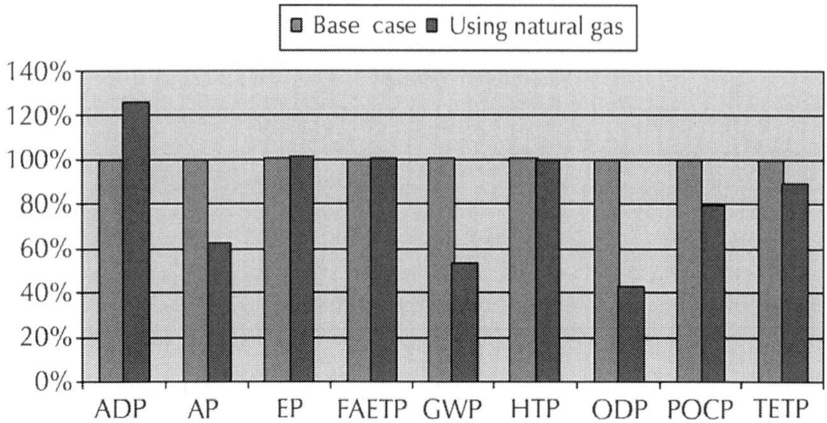

Figure 3: Comparison of environmental impact potentials between the base case using diesel and an assumed case using natural gas for cocoa processing.

From Fig. 3, it is evident that the exclusive use of natural gas instead of diesel in the cocoa processing factory shows many reductions in environmental impact potentials. The reductions are acidification potential (38%), global warming potential (47%), photochemical ozone creation potential (21%) and terrestrial eco-toxicity potential (10%). However, natural gas consumption contributes to energy use and depletion of abiotic resources. This comes from the relatively higher amount of gas required to produce the same amount of energy compared to diesel. Figs. 4 and 5 show that reduction in fertilizer and pesticide use reduces their toxicity impacts significantly. Exclusive use of compost instead of inorganic fertilizers was also tested and the results are shown in Fig. 6. Using compost ensures significant reductions in the environmental impacts of human toxicity and terrestrial eco-toxicity potentials.

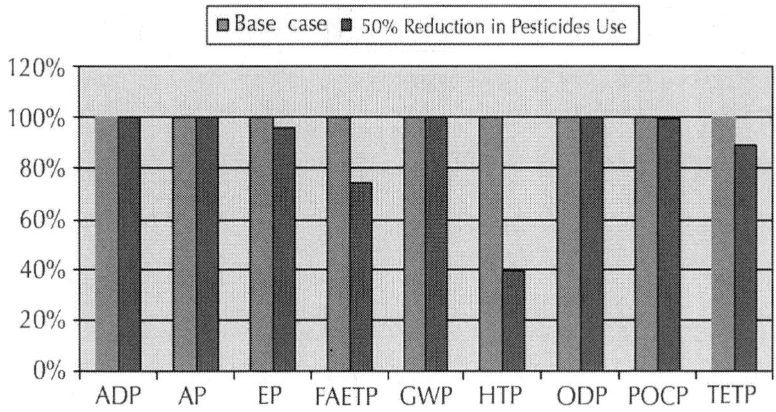

Figure 4: Comparison of environmental impact potentials using pesticides: 100% (base case) versus 50% reduction.

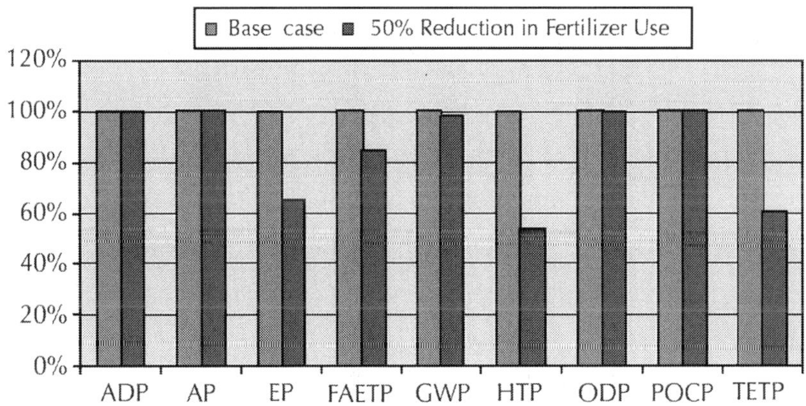

Figure 5: Comparison of environmental impact potentials using fertilizer: 100% (base case) versus 50% reduction.

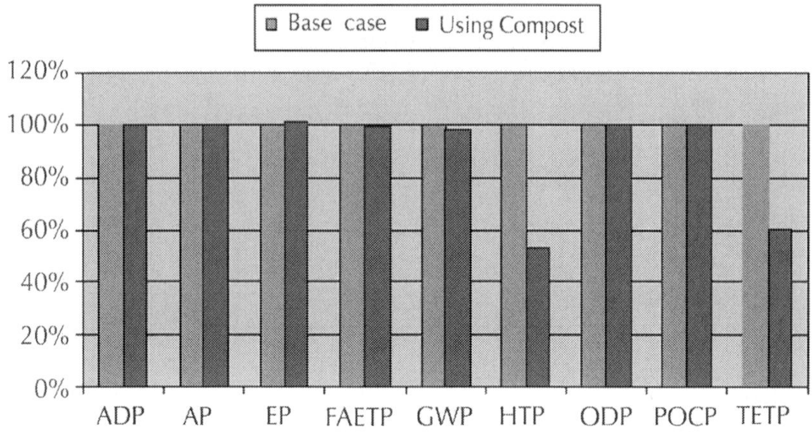

Figure 6: Comparison of environmental impact potentials using fertilizers (base case) versus using compost.

CONCLUSIONS AND RECOMMENDATIONS

The overall aim of conducting this LCA was to measure the potential environmental impacts associated with the production and processing of cocoa in Ghana. This was done in order to establish a scientific basis for improvement analysis towards the sustainability of the production chain. A significant number of impact categories were covered. However, some relevant environmental impacts resulting from cocoa production such as loss of biodiversity and disposal impacts of pod husks have not been assessed and included in the impact assessment, since their methodology is not included in the impact assessment method adopted.

High quality and representative data are critical for reliable LCA results [16]. In this study, data pertaining to electricity generation, fuels and agrochemicals production and transportation were taken from European databases, since specific local or regional databases are lacking. Mungkung et al. [17] have also observed that in developing countries, baseline data, especially describing background systems, are not always available and thus LCA practitioners have to supplement the missing data by using the databases provided in commercial LCA softwares. This sometimes adds to the low confidence level of LCA results. Lack of awareness of the LCA approach in the country was also seen as a barrier. A lot of educational effort had to be made to introduce and explain the concept of life cycle assessment to the people contacted during data collection.

Considering the potential of the method for sustainable development, there is the need to further develop and promote its application in the country's development priority areas. This will require a substantial effort to develop a simplified language that communicates the concepts, tools and benefits of LCA to policy and decision-makers and the development of database relevant to domestic conditions [18]. Current LCA methodologies address only environmental aspects and impacts, therefore recommendations based on LCA's fail to address possible trade-offs between environmental protection and both social and economic concerns in the product life cycle [19,20]. This raises questions about LCA's ability to support actual decision-making in companies, which aim for sustainability, and it creates an incentive for developing LCA methodology to include these other dimensions of sustainability [20].

Despite these general limitations regarding the application of the LCA tool in developing countries like Ghana, one cannot underestimate the environmental perspective offered by a method which makes it possible to identify key environmental issues in support of sustainability measures. For this case study, the LCA tool proved to be successful in identifying and quantifying a number of significant impacts associated with cocoa production and processing in Ghana.

This study is believed to be one of the first LCA studies, and certainly the first in the cocoa industry, in Ghana. The cocoa industry was chosen due to its significant position in the Ghanaian economy. Though the use of the pesticides and fertilizer are assumed to be within acceptable limits [21], this study has revealed that they are a major contributor to the impacts in the cocoa production stage. This means that the government's current Cocoa Disease and Pest Control (CODAPEC) programme must be reviewed, from the environmental point of view. The use of low input systems which rely on integrated pest management involving a high degree of biological control of the major pests and diseases, adequate soil fertility management, as well as high yielding and pests resistant cocoa varieties developed by the CRIG, are recommended to further enhance the environmental friendliness of cocoa production in Ghana [22]. At the industrial processing stage, the use of natural gas instead of diesel oil for roasters and boilers is recommended due to its relatively low emissions. Future study could use other tools to assess land use impacts such loss of biodiversity and soil fertility, etc., to complement the findings made in this LCA study.

ACKNOWLEDGEMENTS

The UNEP/SETAC Life Cycle Initiative awarded the GaBi 4 LCA software used for the project on which this paper is based. The authors would like to thank the following people for their contributions: Frank Asante (Cocoa Processing Company, Tema), Gyima Gyamfi (Cocoa Swollen Shoot Virus Disease Control Division of COCOBOD, Kumasi), Kofi Acheampong and James Buabeng (CRIG).

REFERENCES

1. Appiah MR. Impact of cocoa research innovations on poverty alleviation in Ghana. Ghana Academy of Arts and Sciences Publication; 2004.
2. COCOBOD. Socio-economic study. Final report [MASDAR]. Ghana: Cocoa Board; 1998.
3. Awua PK. The success story of cocoa processing and chocolate manufacturing in Ghana e the success story that demolished a myth. Saffron Walden, Essex: David Jamieson & Associates Ltd.; 2002.
4. ICCO. International cocoa organization 2003/2004. Annual report; 2004. <http://www.icco.org>.
5. Cowell SJ, Clift R. Impact assessment for LCAs involving agricultural production. Int J LCA 1997;2(2):99e103.
6. ISO 14040. Environmental management e life cycle assessment e principles and framework; 1997.
7. ISO 14041. Environmental management e life cycle assessment e goal and scope definition and inventory analysis; 1999.

8. ISO 14042. Environmental management e life cycle assessment e life cycle impact assessment; 2000.
9. ISO 14043. Environmental management e life cycle assessment e life cycle interpretation; 2000.
10. Hauschild M. Estimating pesticide emissions for LCA of agricultural products. In: Weidema BP, Meeusen MJG, editors. Agricultural data for life cycle assessment, vol. II. The Hague: Agricultural Economics Research Institute; 2000.
11. Heathwaite L. Flows of phosphorous in the environment: identifying pathways of loss from agricultural land. In: Weidema BP, Meeusen MJG, editors. Agricultural data for life cycle assessment, vol. 2. The Hague: Agricultural Economics Research Institute; 2000.
12. Adu-Amankwa B, Twumasi K. Feasibility report on pilot plants to process cocoa by-products in Ghana; 2002 [unpublished].
13. Figueira A, Janick J, BeMiller JN. New products from Theobroma cacao: seed pulp and pod gum. In: Janick J, Simon JE, editors. New crops. New York: Wiley; 1993. p. 475e8.
14. Narayanaswamy V, Altham J, Van Berkel R, McGregor M. A primer on environmental life cycle assessment (LCA) for Australian grains. Curtin University of Technology, http://cleanerproduction.curtin.edu.au; 2002.
15. Kooistra KJ, Pyburn R, Termorshuizen AJ. The sustainability of cotton. Consequences for man and environment. Report 223. Science Shop Wageningen University & Research Centre; 2006. ISBN: 90-6754-90-8585- 000-2.
16. UNEP. LCA, what it is and how to do it. Paris, France: UNEP Publications, http://www.unepie.org; 1996.
17. Mungkung R, Udo de haes HA, Clift R. Potentials and limitations of life cycle assessment in setting ecolabelling criteria: a case study of Thai shrimp aquaculture product. Int J LCA 2006;11(1):55e9.
18. Ramjeawon T, Von Blottnitz H, Kituyi E, Mebratu D. LCA knowledge network in Africa. Int J LCA 2005;10(6):449.
19. Guine´e JB, editor. Handbook on life cycle assessment: operational guide to the ISO standards. The Netherlands: Kluwer Academic Publishers; 2001.

20. Dreyer LC, Hauschild MZ, Schierbeck J. A framework for social life cycle impact assessment. Int J LCA 2006;11(2):88e97.
21. Ghana Cocoa Farmers Newspaper. (001); mid-crop 2006.
22. Padi B, Owusu GK. Towards an integrated pest management for sustainable cocoa production in Ghana. In: Proceedings from 1st sustainable cocoa workshop, March 29eApril 3, 1998, Panama; 1998. p. 1, 7e15.

Chapter 14

Cocoa Polyphenols and Their Potential Benefits for Human Health

I. Andújar, M. C. Recio, R. M. Giner, and J. L. Ríos

Departament de Farmacologia, Facultat de Farmàcia, Universitat de València, Avenida Vicent Andrés Estellés s/n, 46100 Burjassot, Valencia, Spain

ABSTRACT

This paper compiles the beneficial effects of cocoa polyphenols on human health, especially with regard to cardiovascular and inflammatory diseases, metabolic disorders, and cancer prevention.

Their antioxidant properties may be responsible for many of their pharmacological effects, including the inhibition of lipid peroxidation and the protection of LDL-cholesterol against oxidation, and increase resistance to oxidative stress. The phenolics from cocoa also modify the glycemic response and the lipid profile, decreasing platelet function and inflammation along with diastolic and systolic arterial pressures, which, taken together, may reduce the risk of cardiovascular mortality. Cocoa polyphenols can also modulate intestinal inflammation through the reduction of neutrophil infiltration and expression of different transcription factors, which leads to decreases in the production of proinflammatory enzymes and cytokines. The phenolics from cocoa may thus protect against diseases in which oxidative stress is implicated as a causal or contributing factor, such as cancer. They also have antiproliferative, antimutagenic, and chemoprotective effects, in addition to their anticariogenic effects.

INTRODUCTION

Plant phenols from numerous plant species are being actively studied as potential treatments for various metabolic and cardiovascular diseases. For example, resveratrol from red wine [1,2], epigallocathechin-3-gallate from green tea [3–5], curcumin from turmeric [6], and quercetin [7–12] from different sources have all been studied as potential therapeutic agents, to induce weight loss, lower blood pressure, attenuate glucose levels and insulin resistance (resveratrol), and improve hemoglobin A1c and lipid profile in humans (epigallocathechin-3-gallate). Other studies also carried out in humans have shown the beneficial effects of grape seeds, chokeberries, coffee, carob, and cocoa [13]. This paper reviews the most recent research on this last substance and its potential benefits for human health.

Since the seventeenth century, cocoa and chocolate have been described as potential medicines. In Europe, various historical documents refer to chocolate's medicinal value; thus, although it was drunk as a beverage with a pleasurable taste, it was primarily eaten as a food to treat a number of disorders, including angina and heart pain [14]. Today, *Theobroma cacao* L. (Sterculiaceae) and its products are consumed worldwide and are studied mainly because of the antioxidant and antiradical properties in vitro of some of their polyphenolic constituents, specially procyanidins and flavan-3-ols [15]. Many studies have described cocoa phenolics as being bioactive compounds, especially prominent for their metabolic and cardiovascular effects. These effects are due, in part, to the antioxidant [16] and antiradical properties of cocoa phenolics [17], which increase the plasma level of antioxidants to prevent the oxidation of LDL-cholesterol [18]. Along with their known antiplatelet effects [19], these particular properties are related to the protective mechanism of cocoa phenolics in heart disease [18]. Other relevant properties include their ability to modulate the immune response [20–24] and their anti-inflammatory [25–27] and anticarcinogenic properties [26, 28]. In this paper, we have focused on the beneficial effects of cocoa polyphenols on human health, especially with regard to cardiovascular and inflammatory diseases, metabolic disorders, and cancer prevention as well as on their antioxidant properties.

POLYPHENOLS CONTENT IN COCOA

Cocoa contains about 380 known chemicals, 10 of which are psychoactive compounds. In their natural state, cocoa beans are virtually inedible because of their high concentration of polyphenols, which gives them an extremely bitter flavor. In a final cocoa product such as chocolate, polyphenol content might decrease from 100% to 10% throughout the different manufacturing processes [29]. For this reason, various authors have focused on the effects of polyphenol-enriched extracts from cocoa [27, 30–32].

Wollgast and Anklam [15] reviewed the polyphenol content of cocoa depending on its origin and the manufacturing process of the final product. Three groups of polyphenols can be identified in cocoa beans: catechins, which constitute about 37% of the polyphenol content in the beans, anthocyanidins (about 4%), and proanthocyanidins (about 58%). Of the catechins, (−)-epicatechin is the most abundant (up to 35%), while (+)-catechin, (+)-gallocatechin, and (−)-epigallocatechin are present in smaller quantities (Figure 1). In the case of anthocyanidins, the main compounds are cyanidin-3-α-L-arabinoside and cyanidin-3-β-D-galactoside, while in the case of procyanidins, the main compounds are dimers, trimers, or oligomers of flavan-3,4-diol linked by 4®8 or 4®8 bonds. Of these, the main procyanidins are B1, B2, B3, B4, B5, C1, and D. The proportions of these derivatives usually change during the manufacturing process. For example, in the dried, fat-free mass of fresh cocoa beans, the total soluble polyphenol content is 15–20%, which is equivalent to 6% in air-dried, nondefatted cocoa beans. In contrast, it is only about 5% in fermented, nondefatted beans because during the fermentation process epicatechin and soluble polyphenol content is reduced by approximately 10 to 20%, anthocyanidins disappear, and procyanidins decrease 3- to 5-fold. There is also a great amount of variation in the phenolic content of cocoa depending on its origin. For example, (−)-catechin concentration in fermented, defatted cocoa beans varies from 16.52 mg/g in Costa Rican cocoa to 2.66 mg/g in cocoa grown in Jamaica.

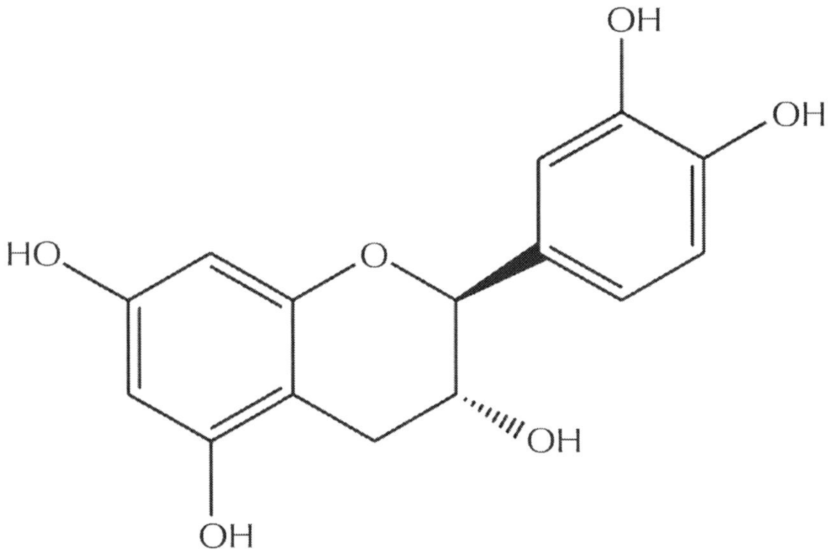

(-)-catechin

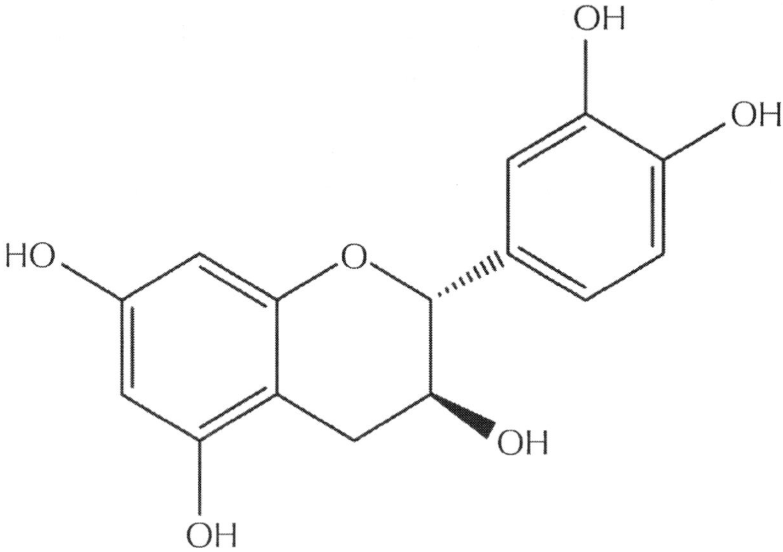

(+)-catechin

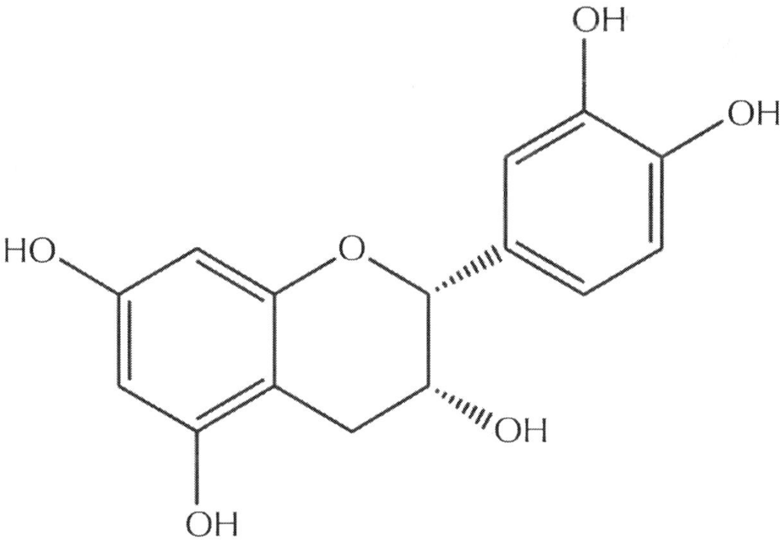

(-)-epicatechin

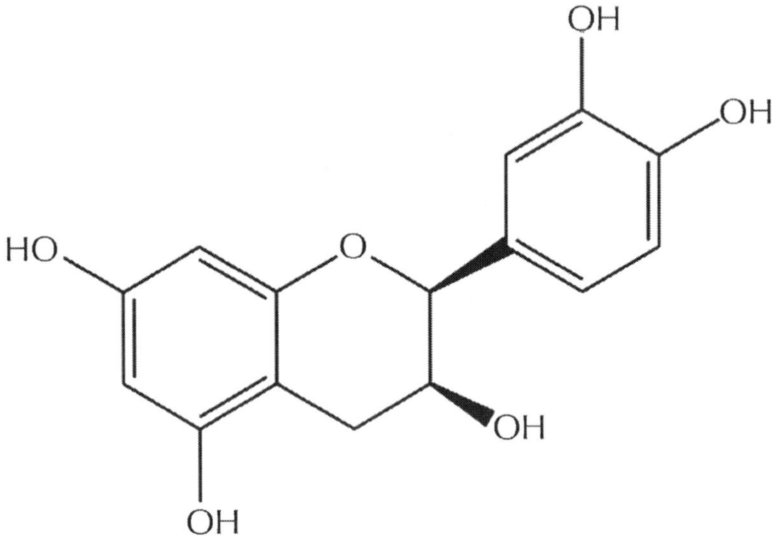

(+)-epicatechin

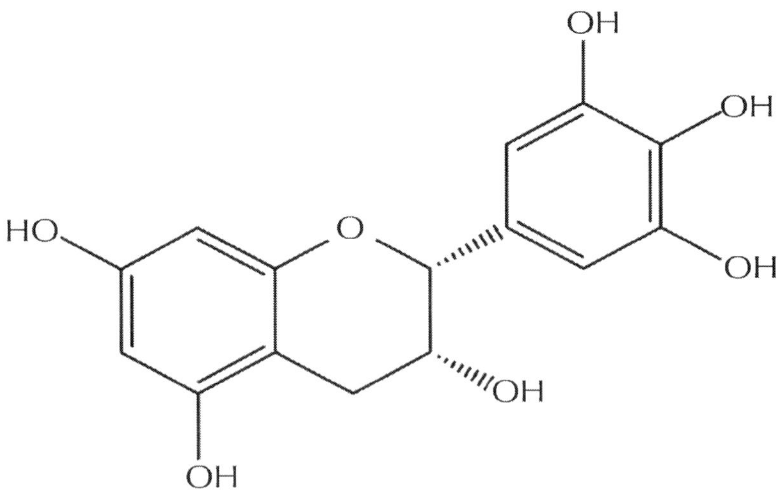

(+)-epigallocatechin

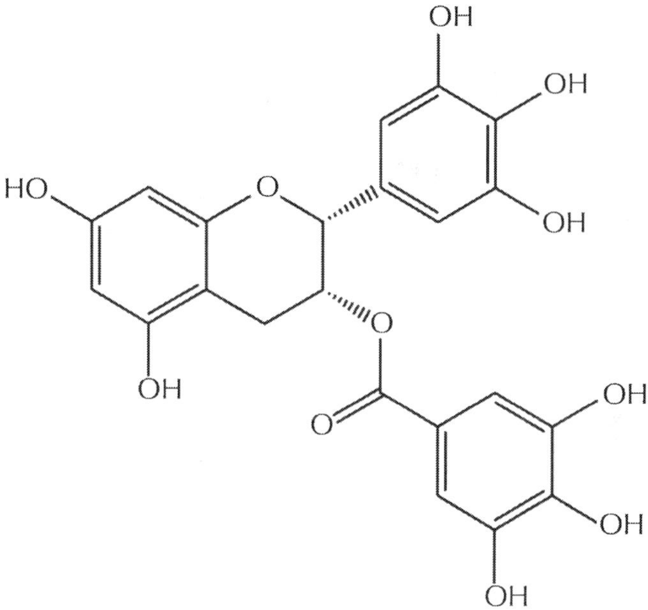

(-)-epigallocatechin gallate

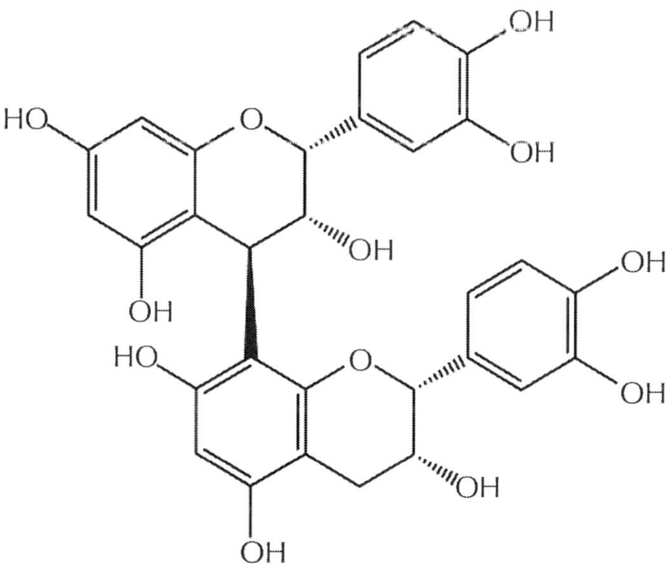

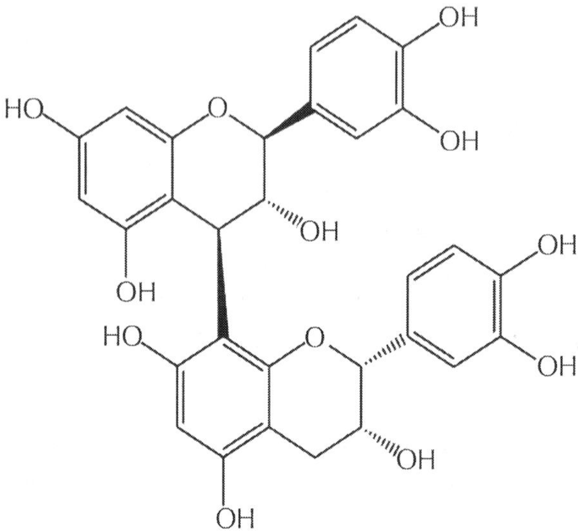

Procyanidin B3

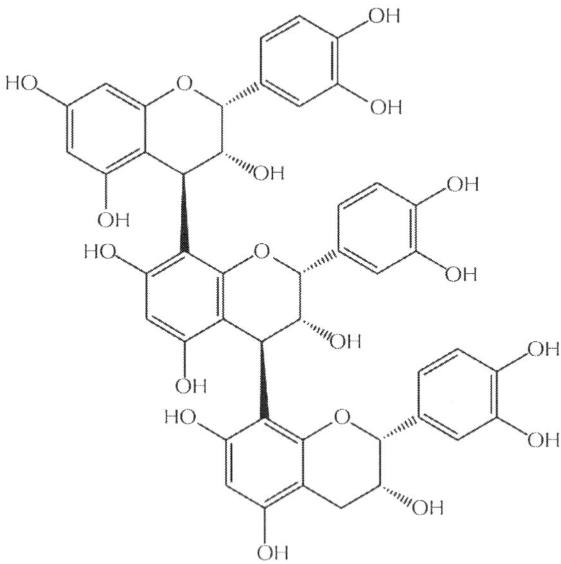

Procyanidin C1

Figure 1: Chemical structure of major cocoa phenolics.

Bioavailability of Phenolics from Cocoa

There is a great controversy surrounding the bioavailability of phenolics in general and of cocoa derivatives in particular. For example, epicatechin is well absorbed, with a maximum plasma concentration at around 2 h and with approximately 20% of consumed epicatechin being excreted in the urine [33]. In the case of dimers, their presence has been also described in plasma. For example, Holt et al. [34] described the presence of procyanidin dimers, especially B2 (epicatechin-(4β-8)-epicatechin) in peripheral blood of healthy adults who had consumed a cocoa beverage.

Gastric Degradation of Cocoa Phenolics

There have been contradictory findings concerning the gastric stability of phenols from cocoa. For example, Spencer et al. [35] reported that, in the acidic environment of the gastric milieu, procyanidin oligomers are hydrolyzed to mixtures of epicatechin monomers and dimers. However, Rios et al. [36] observed that, during gastric transit, procyanidins were remarkably stable in the stomach environment and reached the small intestine intact, available for absorption or metabolism. In addition, their limited absorption in the small intestine could influence the digestive process or the physiology of the gut through direct interaction with the gut mucosa and gut lumen solutes. Consequently, some of the health effects of procyanidins may be associated with the formation by microflora of low-molecular-weight metabolites once the procyanidins reach the colon. This hypothesis has been corroborated by Gonthier et al. [37], who demonstrated that while proanthocyanidins of high molecular weight are not present in plasma or urine, the monomers and aromatic acids produced during their metabolism by intestinal microflora can be found in the latter. These researchers also demonstrated that catechin is well absorbed in the small intestine, with recoveries of 20–40%.

Intestinal Absorption of Phenols from Cocoa

Ortega et al. [38] coupled an in vitro digestion method with a continuous-flow dialyzed duodenal step to construct a model for examining the digestibility and bioaccessibility of polyphenols in cocoa food matrices. Their results show that the fat content of the cocoa samples tested enhances the digestibility of some phenolic compounds, especially procyanidins, during duodenal digestion. The possible mechanism is probably related to the fat fraction's ability to interact with certain phenol compounds following a better micellization of the digested phenols.

A human bioavailability trial with chocolate polyphenols using 40 g of dark chocolate (892 mg of total polyphenols expressed as gallic acid equivalent and 82 mg of epicatechin) showed that epicatechin was indeed absorbed into the blood. The compound was detectable in plasma as glucuronide-conjugated metabolites and sulfate groups after 2 h and was still present after 8 h. The clearance of epicatechin from the plasma compartment was very fast, with an elimination $t_{1/2}$ of 1.9 and 2.3 h for 40 and 80 g of chocolate, respectively [29].

Manach et al. [39] found that the metabolism of simple phenols occurs via a common pathway, being absorbed in the small intestine, and oligomers larger than trimers are unlikely to be absorbed in the small intestine in their native forms. In vitro experiments using single layers of Caco-2 cells as a model of absorption in the small intestine have corroborated the in vivo studies because they showed that only the dimers and trimers of flavan-3-ols are able to cross the intestinal epithelium; thus, procyanidin B2 is poorly absorbed in rats whereas procyanidin B3 is not absorbed at all [39].

In a study on the bioavailability of cocoa polyphenols in healthy volunteers, Tomás-Barberán et al. [30] demonstrated that the flavonoid composition (mainly flavanol monomers and dimers) of cocoa powder can be enhanced through improved processing of the raw material and that these flavanol-enriched samples lead to increases in the cocoa flavan-3-ols present in plasma and urine, thus increasing their bioavailability.

ANTIOXIDANT PROPERTIES OF COCOA PHENOLICS

Scavenging Activity

Cocoa has more phenolics and higher antioxidant capacity than green tea, black tea, or red wine. In fact, Lee et al. [40] found that cocoa contains 611 mg/serving of gallic acid equivalents and 564 mg/serving of epicatechin equivalents. The values for gallic acid equivalents were 1.8, 3.7, and 4.9 higher than those obtained for red wine, green tea, and black tea, respectively, while the epicatechin equivalents were 3.5, 12.0, and 16.6, respectively, for the same beverages. The values for antioxidant activity, measured with the aid of 2,2′-azino-bis(3-ethylbenzthiazoline-6-sulphonic acid) ($ABTS^+$) and 2,2-diphenyl-1-pycrylhydrazyl (DPPH) methods, were 1128 and 836 mg/serving, respectively, expressed as vitamin C equivalent antioxidant capacity. When compared against that of red wine, green tea, or black tea, the relative activity of cocoa was 1.8, 2.5, and 4.4 times higher, respectively, in the $ABTS^+$ test, and 1.7, 2.4, and 4.3 times higher, respectively, in the DPPH test. The authors thus concluded that on a perserving basis, cocoa has higher flavonoid content and antioxidant capacity than red wine (2 times), green tea (2-3 times), and black tea (4-5 times), respectively.

Schinella et al. [16] studied the antioxidant activity of different cocoa extracts and demonstrated their effects on different in vitro experimental models, proving that the antioxidant effects cannot be fully characterized with one single method. The extracts' scavenging capacity was measured as the ability to bleach the stable radicals DPPH and $ABTS^+$, their reducing ability was assessed with the ferric reducing-antioxidant power (FRAP) assay, and their scavenging properties against H_2O_2, HClO, and peroxynitrite were also evaluated. The extracts acted both as a free radical scavenger (DPPH, $ABTS^+$, superoxide, hypochloride, and peroxinitrite) and a lipid peroxidation inhibitor way that was directly proportional to their polyphenol content. The extracts also inhibited superoxide production by polymorphonuclear leukocytes. These findings indicate that cocoa extracts may modify the biological functions of these cells, which are implicated in the initiation and maintenance of inflammation. The extracts also showed scavenging activity against peroxynitrite, both in the presence and in the absence of a physiological concentration of bicarbonate. Under physiological conditions, bicarbonate levels are usually high and significant amounts of nitrosoperoxycarbonate anion adduct can be produced; however, in inflamed tissues where the pH value is lower, different reactions may occur independent of the presence of CO_2. Because peroxynitrite anion can attack a wide range of biological molecules, its inhibition can provide protection against the damage induced by this free radical.

In a previous study, Schroeder et al. [41] examined the effect of a specific cocoa flavanol, (−)-epicatechin in a cellular model (murine aortic endothelial cells) for examining its protective action against peroxynitrite effects, and found that while it effectively blocked the nitration of protein tyrosine residues by peroxynitrite (IC_{50}value = approximately 0.02 mol epicatechin/mol peroxynitrite), its ability to block the oxidative inactivation of glyceraldehyde-3-phosphate dehydrogenase or soybean lipoxygenase (LOX)-1 was not significant (IC_{50} >1 mol epicatechin/mol peroxynitrite).

Lipid Peroxidation

Schinella et al. [16] evaluated the ability of cocoa powder to bring about lipid peroxidation in brain homogenates and human plasma using nonenzymatic generation systems. Cocoa powder inhibited lipid peroxidation in both the homogenates and the plasma but had no effect on the hydroxyl radical, nor did it exhibit ferrous chelating activity. Metal chelating capacity reduces the concentration of the catalyzing transition metal in lipid peroxidation. It has previously been reported that the chelation efficiency of some compounds with a phenyl group on Fe^{2+} depends on the number of hydroxyl groups on the benzene ring and that hydroxyl substitution in the ortho position results in a higher chelating effect. However, because the polyphenol-rich cocoa extract showed neither ferrous nor copper chelating activity, it most likely acts as a chain-breaking antioxidant.

Verstraeten et al. [42] studied the antioxidant and membrane effects of procyanidin B2 (dimer) and procyanidin C1 (trimer) from cocoa in phosphatidyl choline liposomes. Both procyanidins inhibited lipid oxidation of liposomes in a concentration (0.33–5 μM)-dependent manner; moreover, both compounds also increased the membrane surface potential. They did this by decreasing membrane fluidity, an effect that extended into the hydrophobic region of the bilayer, protecting the lipid bilayer from disruption by Triton X-100. Procyanidins can thus interact with membrane phospholipids, probably through their polar head group, to provide protection against the attack of oxidants and other molecules that challenge the integrity of the bilayer.

Resistance to Oxidative Stress

Martorell et al. [43] used model organisms to study the effect of cocoa polyphenols on resistance to oxidative stress. A protocol of acute oxidative stress was optimized in order to test the best conditions to induce oxidative damage in *Saccharomyces cerevisiae*: doses of 4 mg/mL of a polyphenol-enriched cocoa powder led to an increased resistance to hydrogen peroxide (2 and 3 mM) tolerance, similar to those observed in cells treated with vitamin C. To discard the possibility that this was a strain-related effect, the authors repeated the test with other strains belonging to *S. cerevisiae* species. To discover the metabolic target with which the polyphenol-enriched cocoa powder might interact, authors evaluated the transcriptional profile concluding that the protection upon oxidative stress is mediated at least by sirtuins in S. cerevisiae. In order to clarify the function of sirtuin proteins as metabolic targets of cocoa flavonoids, different trials were performed with a more complex model, the nematode *Caenorhabditis elegans*. Worms fed with the cocoa powder were more resistant toward the oxidative stressor hydrogen peroxide, increasing their survival rate from 36% to 53% and their lifespan in a 17% after being fed with the cocoa powder. However, mutant strains of the worm for sirtuin and for the transcription factor DAF-1 did not show resistance to oxidative stress nor prolonged lifespan, suggesting that the effects of cocoa powder are mediated at least through these genes. This work highlights the potential of polyphenol-enriched cocoa powder as a functional ingredient with resistance to oxidative stress.

In a clinical study, Spadafranca et al. [44] investigated the effects of both dark chocolate (860 mg of polyphenols, of which 58 mg were epicatechin) and white chocolate (5 mg of polyphenols, undetectable epicatechin) on plasma epicatechin levels, DNA damage to mononuclear blood cells, and total plasma antioxidant activity in 20 healthy subjects who followed a balanced diet for 4 weeks. Detectable epicatechin levels were only observed 2 h after dark chocolate intake, at which timeless DNA damage to mononuclear blood cells was likewise observed, but both effects were no longer evident after 22 h. No effect on total plasma antioxidant activity was noted. These results demonstrate that dark chocolate improves DNA resistance to oxidative stress in healthy subjects but that this effect is transient, probably due to flavonoid kinetics. Because oxidative stress not only affects both the inflammatory process and atherogenesis, but also has mutagenic power over the human genome, the protective effect of cocoa polyphenols may have important consequences for oncogene expression and cancer pathogenesis.

PHENOLICS FROM COCOA IN METABOLIC AND ENDOCRINE DISORDERS

Cocoa extracts and cocoa procyanidins are potent inhibitors of key enzymes in the digestion of carbohydrates and lipids; these inhibitory activities are related not only to their polyphenol content, but also to the degree of polymerization of cocoa procyanidins, all of which are dramatically affected by the cocoa processing. For example, Gu et al. [45] assessed their inhibitory effects of different cocoa extracts against pancreatic α-amylase, pancreatic lipase, and secreted phospholipase A_2 (PLA_2) in vitro and characterized the kinetics of this inhibition and demonstrated that cocoa extracts and the procyanidins found in them dose-dependently inhibited all three enzymes. They also observed that the inhibitory potency of cocoa procyanidins increases as a function of their degree of polymerization, but they did not examine whether cocoa extracts and/or cocoa procyanidins can inhibit digestive enzymes in vivo and affect related downstream pathways such as aberrant eicosanoid metabolism at dose levels achievable in the diet.

The principal aspects to be treated in this section are the effects of cocoa polyphenols on carbohydrate and lipoprotein metabolism.

Effects on Carbohydrate Metabolism

Hanhineva et al. [46] published an interesting review compiling the effect of different polyphenols, including flavonoids and proanthocyanidins, on carbohydrate metabolism at many levels. In animal models and in a limited number of human studies, polyphenols have been found to attenuate postprandial glycemic responses and fasting hyperglycemia while improving acute insulin secretion and insulin sensitivity. The possible mechanisms of action include inhibition of carbohydrate digestion and glucose absorption in the intestine, stimulation of insulin secretion from the pancreatic β-cells, modulation of glucose release from the liver, activation of insulin receptors and glucose uptake in the insulin-sensitive tissues, and modulation of intracellular signaling pathways and gene expression. These positive effects on glucose homeostasis are also supported by epidemiological evidence on polyphenol-rich diets.

Grassi et al. [47] compared the effects of flavanol-rich dark chocolate and flavanol-free white chocolate with regard to endothelial function, insulin sensitivity, β-cell function, and blood pressure in a randomized crossover trial with 19 hypertensive patients with impaired glucose tolerance. The group of patients treated with flavanol-rich dark chocolate (100 g/day for 15 days) showed decreased insulin resistance and increased insulin sensitivity and β-cell function. Moreover, subjects taking flavanol-rich dark chocolate experienced decreases in systolic and diastolic blood pressure, increased flow-mediated dilation, and decreased total cholesterol and LDL-cholesterol with respect to the control group [47].

Similar effects were described by other authors when reviewing systematically the effect of cocoa polyphenols on cardiovascular risk factors. Shrime et al. [48] did a systematic review and meta-analysis of randomized, controlled trials to evaluate the effect of a flavonoid-rich cocoa fraction on cardiovascular risk factors and showed that the cocoa phenolics decreased insulin resistance and increased flow-mediated vascular dilation, with a maximum effect observed at a flavonoid dose of 500 mg/d.

Hooper et al. [49] in a systematic review of 42 acute or short-term chronic randomized controlled trials including 1297 patients demonstrated a significant reduction of different negative biomarkers of cardiovascular disease, such as insulin resistance. Some parameters were improved in patients after chronic and acute intake of chocolate or cocoa, with significant reductions in serum insulin, and in flow-mediated dilatation. In addition, the reductions in diastolic blood pressure and mean arterial pressure were observed, as well as a slight but significant reduction of LDL-cholesterol and an increase of HDL-cholesterol levels. These effects will be reviewed in the following sections.

Effects on Lipoprotein Metabolism

Cocoa phenolics increase apolipoprotein (Apo) A1 and decrease ApoB production in both the human hepatoma cell line HepG2 and the human intestinal cell line Caco2. Cells incubated for 24 h in the presence of cocoa polyphenols such as (−)-epicatechin, (+)-catechin, procyanidin B2, procyanidin C1, and cinnamtannin A2 at 10 μM increased both ApoA1 protein levels and mRNA expression, even though ApoB protein and mRNA expression decreased slightly. These effects were the result of the upregulation of the mature form of sterol regulatory element binding proteins (SREBPs) and increased LDL receptor activity induced by the cocoa polyphenols. As cholesterol metabolism is known to be regulated by several different mechanisms, it was hypothesized that cocoa polyphenols, like other plant polyphenols, may act via multiple pathways as a regulatory receptor agonist or ligand. The mechanism responsible for the increase in high-density lipoprotein (HDL)-cholesterol associated with daily cocoa intake was also identified [50]. However, other authors have obtained somewhat contradictory results in clinical trials. Khan et al. [51] had previously evaluated the effects of chronic cocoa consumption (42 g/day for 4 weeks) on lipid profile, oxidized LDL particles, and plasma antioxidant vitamin concentrations in 42 high-risk volunteers in a randomized, crossover feeding trial. The analysis of 24 h urine revealed significant increases of phase II metabolites, including glucuronide and sulfate conjugates. The concentration of HDL-cholesterol was significantly higher after cocoa intake in comparison with the control group, which received only milk, while the oxidized LDL levels decreased. No significant changes were registered for total plasma cholesterol, triglycerides, LDL-cholesterol, ApoA1 and ApoB, lipoprotein concentrations, or the LDL/HDL-cholesterol ratio. Other biochemical parameters such as fibrinogen and homocysteine did not vary, and no significant changes were found in the plasma concentrations of vitamins B_1, B_6, B_{12}, C, and E, or in plasma and intraerythrocyte folic acid concentrations. The authors studied the correlation between changes in polyphenols, lipids, and oxidized LDL measurements and observed higher increments in the urinary excretion of cocoa (−)-epicatechin metabolites. Regarding cocoa metabolites derived from intestinal microbial metabolism, an increase in the urinary excretion of 3-hydroxyphenylacetic and vanillic acids was observed in parallel with a significant increase in plasma HDL-cholesterol concentration and a decrease in plasma-oxidized LDL levels.

Jia et al. [52] studied the effect of cocoa on total cholesterol, LDL-cholesterol, and HDL-cholesterol in 8 trials involving 215 participants. Whereas LDL-cholesterol was significantly lowered and total cholesterol marginally lowered, no significant changes in HDL-cholesterol were observed after cocoa consumption. There was no evidence of a dose-effect relationship, of any effect in healthy subjects, or of any change in HDL-cholesterol. The authors concluded that short-term cocoa consumption significantly reduced blood cholesterol, but the changes depended on the amount of cocoa consumed and the health status of participants; indeed, there was no dose response and no effect in healthy participants. To assess the dose-response effect, Almoosawi et al. [53] carried out a randomized, single-blind, crossover study in which 14 overweight/obese subjects were randomized to take either 20 g of dark chocolate with 500 mg of polyphenols and then 20 g of dark chocolate with 1000 mg of polyphenols or vice versa. Participants followed each diet for 2 weeks separated by a 1-week washout period.

Mellor et al. [31] conducted a randomized, placebo-controlled double-blind crossover trial comparing the effects of chocolate with high polyphenol content (16.6 mg of epicatechins in 45 g of chocolate) with those of chocolate with low-polyphenol content (<2 mg of epicatechins in 45 g of chocolate) on lipid profiles, weight, and glycemic control in 12 individuals with type 2 diabetes on stable medication. HDL-cholesterol increased significantly with high polyphenol chocolate, with a decrease in the total cholesterol/HDL ratio and no changes in any of the other parameters under study. The authors concluded that high-polyphenol chocolate is effective in improving the atherosclerotic cholesterol profile in patients with diabetes by increasing HDL-cholesterol and improving the cholesterol/HDL ratio without affecting weight, inflammatory markers, insulin resistance, or glycemic control. However, as stated above, marginally significant effects on LDL- and HDL-cholesterol were found by Hooper et al. [49], although there were no significant effects on total cholesterol, after acute or chronic intake. As a secondary analysis, they assessed the effects on body weight, body mass index, and waist circumference, but few trials reported these outcomes.

Some of these indices were recently evaluated by Shrime et al. [48] who studied the cardioprotective effects of a flavonoid-rich cocoa fraction. The most relevant risk factors evaluated were blood pressure, pulse, total cholesterol, HDL-cholesterol, LDL-cholesterol, triglycerides, body mass index, C-reactive protein, flow-mediated vascular dilation (FMD), fasting glucose, fasting insulin, serum isoprostane, and insulin sensitivity/resistance indices. Results showed that LDL-cholesterol decreased by 0.077 mmol/L, and HDL-cholesterol increased by 0.046 mmol/L, whereas total cholesterol, triglycerides, and C-reactive protein remained the same.

PHENOLICS FROM COCOA IN CARDIOVASCULAR DISEASES

Epidemiological studies demonstrate that regular dietary intake of cocoa polyphenols reduces the risk of coronary heart disease and stroke and is inversely associated with the risk of cardiovascular disease. This effect was described for first time in the Kuna Indians (Panama), who belong to one of the few cultures protected against the age-dependent increase in blood pressure and the development of arterial hypertension. These healthy effects have been correlated with the consumption of high amounts of cocoa. Clinical studies revealed that, compared with other Pan-American citizens, the Kunas have lower blood pressure rates and lower mortality resulting from cardiovascular events [54].

Effects on Arachidonic Acid Metabolism

The consumption of cocoa/chocolate increases plasma antioxidant capacity, diminishes platelet function and inflammation, and decreases diastolic and systolic arterial pressures. Currently available data indicate that daily consumption of cocoa-rich chocolate or polyphenol-enriched cocoa extracts could partially lower cardiovascular disease risk [55, 56]. The mechanism by which the cardiovascular protection of cocoa polyphenols develops is complex, with different mediators and agents being implicated. For example, the effects of cocoa polyphenols on inflammatory mediators such as leukotrienes (LT), nitric oxide (NO) metabolism, and myeloperoxidase- (MPO-) induced modification of LDL indicate that the antioxidant properties of polyphenol-enriched cocoa extracts could be of great interest for developing therapies to prevent cardiovascular disease since oxidative modification of LDL appears to be crucial for atherogenesis. In fact, cocoa polyphenols decrease the plasma concentration of proinflammatory cysteinyl leukotrienes through inhibition of 5-LOX, as demonstrated by Sies et al. [57] in isolated recombinant human 5-LOX (IC_{50} of (–)-epicatechin = 22 μM). Recombinant human platelet 12-LOX, a representative of another subfamily of mammalian lipoxygenases, also was inhibited by (–)-epicatechin (IC_{50} = 15μ M), therefore concluding that this compound and related flavan-3-ols are general inhibitors of mammalian LOX. The inhibition of human 5-LOX by cocoa flavonoids suggests that these compounds exert antileukotriene actions, which may confer a degree of anti-inflammatory, vasoprotective, and antibronchoconstrictory capacity because the principal metabolite of 5-LOX, LTB_4, causes neutrophils to adhere to endothelial cells, thus acting as a potent chemotactic agent for these and other inflammatory cells. It also stimulates the release of lysosomal enzymes and the generation of superoxide anion in neutrophils, while the cysteinyl-LTs increase vascular permeability and contract airway smooth muscle.

Another oxidant enzyme with pronounced effects on inflammation and atherogenesis is MPO, which has also been reported to act as an NO oxidase. In vitro, MPO is capable of binding to LDL, catalyzing oxidative modification and rendering it atherogenic. (–)-epicatechin and procyanidins have been shown to block the MPO/nitrite-mediated and peroxynitrite-mediated lipid peroxidation of LDL; for example, (–)-epicatechin is able to suppress it at concentrations as low as 0.1 μmol/L [58].

Unlike 5-LOX and platelet-type 12-LOX, the reticulocyte-type lipoxygenases (12-LOX and 15-LOX) are catalysts of enzymatic lipid peroxidation. Oxidative modification of LDL by 15-LOX-1, MPO, and peroxynitrite renders its atherogenicity and plays a relevant role in the early developmental stages of atherosclerotic lesions. Agents such as cocoa polyphenols that interfere with the oxidative modification of LDL should therefore be protective for vascular endothelium, including that of coronary arteries.

In conclusion, the inhibitory effects of cocoa polyphenols on 5-LOX and 15-LOX-1 may contribute to their beneficial actions [57].

Effects on NO Bioactivity

Cocoa phenols also promote the bioactivity of NO in vivo, which is critical for protecting the cardiovascular system. Sies et al. [57] compared the effects of cocoa beverages containing different contents of flavanols among selected volunteers, concluding that ingestion of flavanol-rich cocoa by individuals with diminished endothelial function led to an increase in the plasma NO concentration approaching the mean concentrations of healthy control subjects and doubled flow-mediated dilation after 2 h. Plasma concentrations of NO were correlated with flow-mediated dilation, which suggests that the improvement of endothelial function might be attributable to increased bioavailability of NO. In contrast, these changes did not occur among individuals who received either water or a low-flavanol cocoa drink, which excluded effects of beverage constituents other than flavanols. The NO-promoting effects of cocoa polyphenols may also contribute to a decrease in blood pressure because they interfere with NO metabolism at several levels. They especially favor the synthesis of NO and suppress deleterious metabolites such as superoxide, peroxynitrite, and NO_2 [57].

As Corti et al. [54] cite, these compounds also induce NO-dependent vasodilation in the rat aorta and in the finger or forearm circulation of healthy humans or patients with cardiovascular risk factors. Cocoa lowers vascular arginase activity in human endothelial cells, augmenting local levels of L-arginine, which is then transformed into NO by endothelial nitric oxide synthase (eNOS). Once released, NO increases intracellular cyclic guanosine monophosphate (cGMP) concentrations, inducing a relaxation of vascular smooth muscle cells. NO not only leads to vasodilatation, but also prevents leukocyte adhesion and migration, smooth muscle cell proliferation, and platelet adhesion and aggregation. In patients with cardiovascular risk factors, including smoking, a cocoa drink high in flavonol content rapidly enhances the circulating pool of bioactive NO and augments flow-mediated vasodilation. Moreover, chronic consumption of a high-flavanol diet is associated with high urinary excretion of NO metabolites, which is consistent with augmented NO production or diminished degradation. Moreover, flavonoid-enriched cocoa results in regional changes in cerebral blood flow and an overall increased blood flow. In elderly humans on a high-flavanol diet, cerebral blood flow velocity in the middle cerebral artery has been shown to increase, suggesting that cocoa flavanols may protect against dementia and stroke [54]. The improvement of endothelial function is probably one of the mechanisms by which polyphenols confer cardiovascular protection. Indeed, polyphenols are able to induce NO-mediated, endothelium-dependent relaxations in a large number of arteries, including the coronary artery. In some arteries they can also induce endothelium-derived hyperpolarizing factor (EDHF)-mediated relaxations. Taken together, these mechanisms may help explain the antihypertensive and cardioprotective effects of polyphenolsin vivo [58].

EFFECTS OF COCOA PHENOLICS ON OTHER MEDIATORS IMPLICATED IN VASCULAR PROPERTIES

Cocoa procyanidins are potent inhibitors of mitogen-activated protein kinase kinase (MEK) and membrane type-1 (MT1) matrix metalloproteinase (MMP). They subsequently inhibit the expression and activation of pro-MMP-2 as well as the invasion and migration of human vascular smooth muscle cells (VSMCs). In fact, both cocoa procyanidin fraction (3 μg/mL) and procyanidin B2 (5 μM) were found to strongly inhibit the thrombin-induced activation and expression of pro-MMP-2 in VSMCs, the thrombin-induced invasion and migration of VSMCs, MT1-MMP activity, and the thrombin-induced phosphorylation of extracellular signal-regulated protein kinase (ERK) (but not that of MEK) in VSMCs. Kinase and pull-down data show that cocoa procyanidin fraction and procyanidin B2 inhibit MEK1 activity, binding directly with glutathione-S-transferase-MEK1. These results may partially explain the molecular action of cocoa's antiatherosclerotic effects [59].

There is evidence of the structure-activity relationship of endothelium-dependent vasodilator responses to procyanidins in isolated vessels. This is consistent with the hypothesis that these responses are mediated by a procyanidin receptor, which could represent an important target for new agents to treat endothelial dysfunction. Because this response involves signaling through superoxide or hydrogen peroxide, it could even be considered prooxidant. Flow-mediated dilatation also involves signaling through hydrogen peroxide. Hence, these findings bolster the hypothesis that procyanidins trigger an endothelial response similar to that of laminar shear stress [60].

In a controlled trial, Monagas et al. [32] evaluated the effects of chronic cocoa consumption on cellular and serum biomarkers related to atherosclerosis in 42 high-risk patients in a randomized crossover feeding trial. All subjects received 40 g cocoa powder with 500 mL skim milk per day or only 500 mL skim milk per day for 4 weeks. No significant changes in the expression of adhesion molecules on T lymphocyte surfaces were observed. However, in monocytes, the expression of VLA-4, CD40, and CD36 was significantly lower in the cocoa intake group compared to the control group. In addition, serum concentrations of the soluble endothelium-derived adhesion molecule P-selectin and intercellular adhesion molecule (ICAM)-1 were significantly lower after cocoa intake than after milk intake. The authors concluded that the intake of cocoa polyphenols may modulate inflammatory mediators in patients at high risk for cardiovascular disease; indeed, these anti-inflammatory effects may contribute to the overall benefits of cocoa consumption against atherosclerosis.

The vascular actions associated with transforming growth factor (TGF)-β_1 and homeostatic modulation of its production by flavan-3-ols and procyanidins from cocoa may constitute another mechanism by which phenolics from cocoa can potentially benefit cardiovascular health. In this context, Mao et al. [61] studied the effect of flavan-3-ols and procyanidins fractions (monomers through decamers) isolated from cocoa on the secretion of the cytokine TGF- β_1 in resting human peripheral blood mononuclear cells (PBMCs) from 13 healthy subjects. When cells of individuals with low baseline levels of TGF- β_1 were stimulated by individual flavan-3-ol and procyanidin fractions (25 μg /mL), TGF- β_1 release was enhanced in the range of 16–66%. The low-molecular-weight fractions (≤pentamers) were more effective at augmenting TGF- β_1 secretion than their larger counterparts (hexamer or higher), with the monomer and dimer inducing the greatest increases (66% and 68%, resp.). In contrast, in the case high TGF-β_1 baseline subjects, the TGF- β_1 secretion was inhibited, being the inhibition most pronounced for trimmers through decamers (28–42%, resp.) and moderate for monomers to dimers (17–23%, resp.). Given the vascular actions associated with TGF- β_1 release, this homeostatic modulation by cocoa polyphenols offers an additional mechanism by which polyphenol-rich foods can benefit cardiovascular health.

Cocoa Phenols and Cardiovascular Disease

Ding et al. [62] carried out a systematic review of experimental, observational, and clinical studies of the relationship between chocolate and cocoa and the risk of cardiovascular disease. The data from a total of 136 publications suggests that cocoa and chocolate may exert beneficial effects with regard to cardiovascular risk via various mechanisms, including lowering blood pressure, anti-inflammation, anti-platelet function, increasing HDL level while decreasing LDL oxidation. In addition, prospective studies of flavonoids suggest that the phenolic content of chocolate may reduce the risk of cardiovascular mortality.

Phenolics from cocoa also decrease the tendency of blood to clot. For example, an acute dose of flavanols and oligomeric procyanidins from cocoa powder was found to inhibit platelet activation and function for over 6 h in humans. In a blinded parallel-design study, 32 healthy subjects were assigned to consume 234 mg cocoa phenolics a day [63].

EFFECTS ON PLATELET FUNCTION

Platelet function was determined by measuring platelet aggregation, ATP release, and the expression of activation-dependent platelet antigens. After 28 days of treatment, cocoa polyphenol supplementation increased plasma epicatechin and catechin concentrations by 81% and 28%, respectively, while decreasing platelet function. These findings are in agreement with results from studies that used higher doses of cocoa flavanols and procyanidins [63]. Moreover, cocoa decreases not only platelet aggregation, but also adhesion. The effects of cocoa polyphenols could thus be explained in part by their ability to reduce the ADP-induced expression of the activated conformation of glycoprotein IIb/IIIa surface proteins [54]. Various studies have indicated that dark chocolate induces coronary vasodilation, improves coronary vascular function, and decreases platelet adhesion. These immediate beneficial effects are paralleled by a significant reduction in serum oxidative stress and have been positively correlated with changes in serum epicatechin concentration. Flammer et al. [64] studied these effects on coronary vascular and platelet function in 22 heart transplant recipients in a double-blind, randomized study. The ingestion of 40 g of flavonoid-rich dark chocolate (containing 0.27 mg/g of catechin and 0.9 mg/g of epicatechin, with a total polyphenol content of 15.6 mg of epicatechin equivalents per gram) increased coronary artery diameter, improved endothelium-dependent coronary vasomotion, and decreased platelet adhesion, all of which remained unchanged in the control group.

Carnevale et al. [65] carried out a crossover, single blind study with 20 healthy subjects and 20 smokers who received 40 g of dark (cocoa > 85%) or milk (cocoa < 35%) chocolate. Compared to healthy subjects, smokers showed enhancements in platelet recruitment, platelet formation of reactive oxygen species (ROS) and eicosanoids, and nicotinamide adenine dinucleotide phosphate (NADPH)-reduced oxidase (NOX)2 activation. In the dark chocolate group, platelet ROS, 8-iso-prostaglandin (PG)$F_{2\alpha}$, and NOX2 activation decreased significantly, but in the subgroup of healthy subjects taking dark chocolate, platelet variables did not change. No changes were detected in either of the groups treated with milk chocolate. The authors concluded that dark chocolate inhibits platelet function by lowering oxidative stress only in smokers; this effect seems to be dependent on its polyphenol content.

EFFECTS ON ARTERIAL HYPERTENSION

As discussed above, animal and human studies suggest that cocoa and chocolate could regulate NO bioavailability, affecting endothelial function and, in turn, blood pressure. Although these studies are not definitive and many questions remain, the body of evidence suggests that it is worth considering diets that can provide benefits to patients with high blood pressure; indeed, these benefits will be more significant in subjects that do not have their hypertension under control [66].

As mentioned above, significant reductions in diastolic blood pressure and mean arterial pressure after chronic intake were observed in Hooper's review [48].

Cienfuegos-Jovellanos et al. [67] evaluated the effects of various doses of a natural flavonoid-enriched cocoa powder (50–600 mg/kg) on spontaneously hypertensive rats. The polyphenol extract produced an antihypertensive effect in hypertensive rats without modifying the arterial blood pressure in normotensive rats. However, no dose-response effect was observed: the maximum effect on the systolic blood pressure of hypertensive rats was reached with 300 mg/kg of extract, while the maximum effect on the diastolic blood pressure, was achieved with a dose of 100 mg/kg. In a complementary study, Almoosawi et al. [53] found that the 500 mg polyphenol dose was equally effective in reducing systolic blood pressure, and diastolic blood pressure as the 1000 mg polyphenol dose, which suggests that a saturation effect might occur with increased doses of polyphenols.

Akita et al. [68] studied the effect of dietary intake of cocoa polyphenols on heart rate, blood pressure, and plasma lipid concentrations. The area of atherosclerotic lesions in the aortas of the polyphenol-treated group was significantly smaller than in the control group. In addition, whereas the high-frequency power of heart rate variability of the rabbits in the standard diet group decreased significantly with ageing, there were no differences between long- and short-term treatments in the polyphenol group. Moreover, cocoa liquor polyphenols preserved parasympathetic nervous tone, although this decreased significantly with ageing in the standard diet group. No differences in the plasma lipid concentrations were observed. The authors concluded that cocoa liquor polyphenols may play an important role in protecting cardiovascular function.

To date, epidemiological data on cocoa polyphenols has been scarce. Rimbach et al. [69] reviewed different relevant clinical trials and selected the results of five of them in which the following conclusions were established: in a cohort of elderly men, cocoa intake was inversely associated with blood pressure as well as with 15-year cardiovascular and all-cause mortality; a prospective study examined flavonoid intake and cardiovascular disease mortality in postmenopausal women; a meta-analysis of 133 studies on flavonoids, flavonoid-rich foods, and cardiovascular disease risk found a correlation between chocolate intake and increased flow-mediated dilation along with reductions in both systolic and diastolic blood pressure; a cohort study involving over 1100 nondiabetic patients assessed the long-term effects of chocolate consumption amongst patients with established coronary heart disease; finally, a study conducted with patients surviving their first acute myocardial infarction found that chocolate consumption was associated with significant reductions in cardiac mortality and beneficial effects on vascular health. However, these studies raise many doubts. For example, the products used in controlled studies often contain much higher polyphenol contents than most commercially available products. It thus needs to be established whether the consumption of products with lower polyphenol content is associated with any health benefits in humans. Moreover, many of these studies have mainly been funded by industrial sponsors, and it must be remembered that the food industry is encouraged to label the flavonoid content on cocoa-derived products [69].

In a 6-month clinical trial carried out by Ried et al. [70] with 36 prehypertensive healthy adult volunteers, at content levels of 120–139 and 80–89 mmHg, no significant differences were observed in the blood pressures of the treated and control groups. However, a marked difference in acceptability between the two treatment forms (chocolate or capsule) was observed due to the amount of chocolate that had to be eaten (50 g of dark chocolate every day). In contrast, other studies [71] have reported on blood pressure changes in response to flavanol-rich cocoa in healthy subjects as well as in prehypertensive and hypertensive patients, suggesting that the inclusion of moderate amounts of flavanol-rich cocoa or chocolate in the daily diet may potentially ameliorate or delay the onset of hypertension. Indeed, it has been estimated that a 3 mmHg reduction in systolic blood pressure reduces the relative risk of stroke mortality by 8%, of coronary artery disease mortality by 5%, and of all-cause mortality by 4%.

Actis-Goretta et al. [72] examined the effects of different isolated polyphenols from cocoa on angiotensin converting enzyme activity; several procyanidins (dimer and hexamer) and epigallocatechin significantly inhibited the enzyme activity, whereas monomers, flavonols, phenolic acids, and resveratrol at similar concentrations showed no activity. Authors established three levels of IC_{50}: values in the mM range for monomeric flavan-3-ols, in the 100 μM range for dimer and trimer, and in the 10 μM range for larger procyanidins, although these concentrations should be interpreted with caution as they were estimated using nonphysiological in vitro conditions. These in vitro effects were correlated with the in vivo reductions in blood pressure observed in several experimental models [72].

Cherniack [13] reviewed the effects of polyphenols on metabolic syndrome and noted that eight patients with coronary artery disease who were given cocoa drinks containing a 375 mg dose of cocoa polyphenols twice a day for 1 month had improved vasodilatory capacity in their brachial arteries and were able to produce larger numbers of endothelial progenitor cells. Moreover, the systolic blood pressure of subjects taking the larger dose cocoa decreased 4.2 mmHg, an effect which was not observed in the group that consumed only 9 mg twice a day. In addition, meta-analyses of human trials have concluded that cocoa can lower systolic and diastolic blood pressure by 4.5 and 2.5, respectively [13].

A meta-analysis with 10 randomized controlled trials comprising 297 individuals confirmed the blood pressure-lowering capacity of flavanol-rich cocoa products in a larger set of trials than previously reported. Both the systolic and diastolic blood pressures decreased in either healthy normotensive adults or patients with prehypertension/ stage 1 hypertension. However, significant statistical heterogeneity across studies was described [73].

In summary, critical attention must be paid to the flavanol content of cocoa foods since manufacturing processes significantly reduce its concentration in the final product. The high fat and sugar content of many cocoa snacks and confectionaries must also be taken into account. However, even with these limitations, the latest findings on the effects of flavonoid-rich cocoa are of clinical relevance and hold promise for the development of new, low calorie cocoa foods and beverages. Further research is needed to characterize the mechanisms of action leading to the final effects of cocoa. Also, before recommending cocoa products as a treatment option in hypertensive and at-risk patients, well-designed, long-term clinical trials are needed to definitively confirm the putative beneficial role of long-term consumption of flavanol-rich foods. Specific attention should be paid to clarifying questions concerning the most appropriate dose, the best targets, and eventual long-term and high-dose side effect profiles [69].

EFFECTS OF COCOA PHENOLICS ON THE IMMUNE SYSTEM

Kenny et al. [21] demonstrated that cocoa oligomers are potent stimulators of both the innate immune system and early events in adaptive immunity. They supported their hypothesis with evidence that the chain length of flavanol fractions showed a significant effect on cytokine release from both unstimulated and LPS-stimulated PBMCs after 16 h treatment (20 μg/mL). For example, there was a marked increase in LPS-induced synthesis of interleukin (IL)-1β, IL-6, IL-10, and TNF-α in the presence of long-chain flavanol fractions (hexamers to decamers). Both long- and short-chain flavanol fractions stimulated the production of granulocyte macrophage colony-stimulating factor (GM-CSF) in the absence of LPS and increased the expression of the B cell markers CD69 and CD83. However, different results were obtained by other authors using phytohemagglutinin A (PHA)-stimulated PBMCs, in which the production of these cytokines depended on the degree of polymerization. For example, IL-1β was suppressed by monomers to tetramers, but stimulated by pentamers to decamers, and TNF-α was clearly increased only by decamers [25]. These results indicate that some effects of cocoa polyphenols depend on their degree of polymerization as well as the agent used for cell stimulation.

ANTI-INFLAMMATORY PROPERTIES OF PHENOLICS FROM COCOA

Sies et al. [57] reviewed the effects of cocoa phenolics on inflammatory mediators such as leukotrienes. Some of these effects have been previously cited for their implication in cardiovascular diseases. They reported that (−)-epicatechin inhibits the dioxygenase and LTA_2 synthase activities of 5-LOX, the first two consecutive steps in the conversion of arachidonic acid into various proinflammatory agents, including leukotrienes, which are catalyzed by the same enzyme. (−)-Epicatechin has also been shown to inhibit 15-LOX-1. When other cocoa products (e.g., oligomeric procyanidins) were tested as potential inhibitors, the authors observed that only small procyanidins (dimers to pentamers) showed effects similar to those of monomers. However, the pattern against 12-LOX-1 was quite different. The inhibitory potency first decreased from monomers to tetramers but then increased with the oligomer size, with the decamer being the most potent. These effects were also demonstrated in vivo. The subjects with a diet including polyphenol-rich cocoa products had lower levels of plasma leukotrienes LTC_4, LTD_4, and LTE_4, along with increased levels of prostacyclin 2 h after treatment [25].

Effects on Proinflammatory Cytokine Release

Ramiro et al. [74] studied the effects of a cocoa extract on the secretion and RNA expression of various proinflammatory mediators by macrophages. Of these, monocyte chemoattractant protein (MCP)-1 and tumor necrosis factor (TNF)-α were significantly and dose-dependently diminished by the extract. All cocoa flavonoids tested were capable of reducing MCP-1 secretion after 6 h of LPS activation. Thus, epicatechin (100–200 μM, corresponding to 29–58 μg/mL) reduced MCP-1 secretion, exhibiting an inhibition of up to 28%; cocoa extract (50 μg/mL) also diminished MCP-1 secretion, and this effect was significantly stronger than that produced by epicatechin (58 μg/mL). Isoquercitrin (5–40 μM, which represents 2.32–18.6 μg/mL) was the most effective in lowering MCP-1 levels, causing a 28–65% dose-dependent decrease. As for TNF-α, both epicatechin (58 μg/mL) and isoquercitrin (18.6 μg/mL) significantly reduced TNF-α secretion, achieving an inhibition of approximately 35% and 20%, respectively. Cocoa extract (50 μg/mL) also significantly decreased TNF-α secretion by about 15%. These effects may be produced, at least in part, at the transcriptional level. A 58 μg/mL concentration of epicatechin reduced TNF-α and IL-6 mRNA levels, having a higher effect on IL-1αmRNA expression, and one which was very close to basal levels. A 25 μg/mL concentration of cocoa extract produced a greater downregulation of TNF-α, IL-1α, and IL-6 mRNA. Following cocoa extract treatment, IL-1α mRNA remained at basal levels and LPS-induced TNF-α mRNA overexpression was reduced. In addition, the extract decreased NO secretion in a dose-dependent manner and with a greater effect than that produced by epicatechin.

There is experimental and clinical evidence that polyphenols from cocoa may reduce the production of more specific proinflammatory cytokines than those listed above, such as IL-1β and IL-2, while stimulating the production of anti-inflammatory cytokines such as TGF-βand IL-4. Selmi et al. [25] reviewed the studies carried out in resting or phytohemagglutinin A (PHA)-stimulated PBMCs. As they recall, IL-2 production is inhibited by pentamers to heptamers in PHA-stimulated PBMCs at mRNA level. In the case of IL-1β, they describe the biphasic effect of flavonol fractions from monomers to tetramers, which suppressed its mRNA expression and protein secretion, while fractions from pentamers to decamers stimulated IL-1β production. Similar effects were observed in the case of TNF-α, which was increased 4-fold by decameric fractions, while monomer and dimer fractions increased TNF-α production only slightly. In the case of the anti-inflammatory cytokine IL-4, large cocoa flavanols fractions (pentamer through decamer) enhance protein levels from resting PBMCs, whereas the monomers to tetramers were not stimulatory. However, in PHA-stimulated cells, the monomer was the only flavanol fraction able to induce IL-4.

Effects on Nuclear Factor Activation and Signal Transcription

The regulatory effects of cocoa flavanols on nuclear factor-κB (NF-κB) activation have also been studied. (−)-epicatechin, (+)-catechin, and their dimeric forms were found to inhibit the NF-κB activation induced by 12-O-tetradecanoylphorbol-13-acetate (TPA) in T cells, with a clear reduction of NF-κB-DNA binding activity that leads to a reduction in IL-2 production. These effects were related to direct interaction with the inhibitor of κB (IκB) to prevent its phosphorylation, thereby preventing NF-κB activation [25]. In a complementary study, Zhang et al. [75] investigated the effects of procyanidin dimer B2 on COX-2 expression and demonstrated that pretreatment for 30 min of differentiated human monocytic cells (THP-1) in culture with procyanidin B2 (50 μM) reduced the LPS-induced expression of this enzyme. This effect was correlated to the decreased activation of ERK, Jun-terminal kinase

(JNK), and p38 mitogen-activated protein kinase (MAPK), as well as to the suppression of NF-κB activation through stabilization of IκB proteins, which suggests that these signal-transducing enzymes could be potential targets for procyanidin B2.

INFLAMMATORY BOWEL DISEASE (IBD)

Various studies conducted in both in vivo and in vitro models have provided evidence that pure polyphenols and natural polyphenol plant extracts can modulate intestinal inflammation. These studies indicate that pure polyphenols and plant polyphenolic extracts may play an anti-inflammatory role via the modulation of intracellular signaling cascades in intestinal cells. However, the number of studies on this subject is limited and the results have been contradictory [76].

Romier-Crouzet et al. [77] studied different polyphenol extracts, including those from cocoa, and tested their ability to modulate intestinal inflammation using human intestinal Caco-2 cells treated for 4 h and stimulated with IL-1β for 24 or 48 h. Surprisingly, they found that in the absence of any proinflammatory stimulus, the polyphenol extract of cocoa (50 μmol of gallic acid/L) induced a basal PGE_2 synthesis in Caco-2 cells. They attributed this effect to cyclooxygenase (COX)-1 induction. However, the same extract decreased PGE_2 synthesis 4.6- and 2.2-fold in Caco-2 cells stimulated for 24 or 48 h, respectively. As COX-1 seems to be implicated in maintaining mucosal integrity, the induction of COX-1 activity could be a new beneficial property of cocoa polyphenol extract.

A recent study performed by Andújar et al. [27] demonstrated the anti-inflammatory effects of a cocoa polyphenol-enriched extract (500 mg/kg) on an experimental model of IBD. The extract was shown to reduce weight loss, improve normal stool consistency, and reduce or eliminate visible blood in feces in a dextran sulfate sodium (DSS)-induced model of colitis. In parallel, it reduced neutrophil infiltration, NO production, COX-2 expression, and expression of the signal transducers and activators of transcription (STAT)-3 and STAT1α phosphorylation; however, the authors could not demonstrate a clear inhibitory effect on NF-κB in the nuclear extract of the mid and distal colon. The activation of STAT1 triggers an important signaling pathway for many cytokine and growth factor receptors. STAT1 plays a relevant role in IBD and especially in ulcerative colitis, because there is an increased expression and activation of this transcription factor in the neutrophils and monocytes present in the intestinal lamina propria. In addition, STAT3 can be activated by different mediators and after dimerization and migration to the nucleus; it can induce the activation of various genes, with the consequent induction of other proinflammatory mediators. The inhibition of these transcription factors decreases the production of COX-2β, IL-6, IL-1, and TNF-α, which may explain the effect of cocoa polyphenol on experimental models of ulcerative colitis. In isolated peritoneal macrophages, the authors observed reductions in the proinflammatory cytokines IL-1β, IL-6, and TNF-α, probably as a consequence of the inhibition of the transcriptional factors STAT3 and STAT1α. However, interferon (IFN)-γ was not modified. Finally, working with LPS-stimulated RAW 264.7 cells, a reduction of nuclear p65 was observed in the NF-κB reporter gene assay.

The controversy on the effects of cocoa consumption was clarified somewhat by Vázquez-Agell et al. [78], who demonstrated that cocoa polyphenols reduce NF-κB activation in peripheral blood mononuclear cells in humans. They evaluated the effects of 40 g of cocoa powder with either milk or water against those of milk alone in 18 healthy volunteers. NF-κB activation in leukocytes and adhesion molecules was measured before and 6 h after each intervention. The results established that consumption of cocoa with water significantly decreased NF-κB activation from the baseline value. Cocoa with milk produced no change, and milk alone led to a significant increase in NF-κB activation. The concentration of soluble intercellular adhesion molecule (sICAM)-1 significantly decreased 6 h after cocoa-water and cocoa-milk interventions, with E-selectin decreasing only after cocoa-water intervention. No significant changes were observed in the concentration of soluble vascular cell adhesion molecule (sVCAM)-1. The authors concluded that cocoa consumption may confer beneficial anti-inflammatory effects mediated by inhibition of the NF-κB-dependent transcription pathway or by direct interaction with certain cytokines and that the food matrix probably plays a crucial role in the modulation of this effect.

Ingestion or systemic administration of polyphenols may inhibit gene expression and induce phase II antioxidant and detoxifying proteins, which may help them prevent and/or ameliorate inflammatory colitis. Because enteral nutrition plays an important role in the management of IBD, artificial nutrition attempts to safely administer the essential and otherwise beneficial constituents of food to patients with an impaired ability to ingest or digest food; yet polyphenols are not included in the formulae. The addition of these compounds to artificial nutritional formulae could improve the outcome of patients with IBD in need of enteral nutrition [79].

PHENOLICS FROM COCOA IN CANCER PREVENTION

The antioxidant properties of cocoa or cocoa-derived products enriched in flavonoids may help protect against diseases in which oxidative stress is implicated as a causal or contributing factor. The most relevant antioxidant properties have been described above; the following sections are a compilation of these effects and their implications for cancer.

Cell Protection against Active Oxygen Radicals

Cocoa liquor polyphenols could thus be therapeutically beneficial, as they effectively inhibit mitomycin C-induced DNA damage and exert anticlastogenic effects. Because oxidative DNA damage seems to be implicated in mutagenesis and carcinogenesis, the anticlastogenic activities of cocoa extract, both through the inhibitory effect of cocoa polyphenols on the DNA strand cleavage induced by mitomycin C in vitro and through their effect against the formation of micronuclei induced by mitomycin C in bone marrow cells and peripheral blood cells of mice, may explain cocoa's role in the prevention of DNA damage, which probably occurs through the scavenging of active oxygen radicals generated in reactions initiated by mitomycin C [80].

Cho et al. [81] studied the protective effects of a cocoa procyanidin fraction and procyanidin B2 against the apoptosis of PC12 rat pheochromocytoma cells induced by hydrogen peroxide. In this case, the fraction (1 and 5 μg/mL) and procyanidin B2 (1 and 5 μM) reduced cell death by attenuating the hydrogen peroxide-induced fragmentation of the nucleus and DNA in PC12 cells. Hydrogen peroxide acts in part by inducing cleavage of poly(ADP-ribose)polymerase (PARP) and downregulating Bcl-XL and Bcl-2 in PC12 cells, but pretreatment with cocoa procyanidin fraction or procyanidin B2 before hydrogen peroxide treatment diminished PARP cleavage, increased Bcl-XL and Bcl-2 expression, and also inhibited activation of caspase-3 by hydrogen peroxide while attenuating the phosphorylation of JNK and MAPK. Cocoa procyanidin fraction and procyanidin B2 thus have protective effects against both hydrogen peroxide and the oxidative stress induced by ROS, which has been strongly associated with the pathogenesis of neurodegenerative disorders, including Alzheimer's disease. A cocoa-rich diet also exerts antiproliferative effects in azoxymethane-induced colon cancer, decreasing ERK, protein kinase B (Akt), and cyclin D1 levels, as demonstrated in Wistar rats. In addition, it has proapoptotic effects, as evidenced by reduced Bcl-XL levels and increased levels of Bax and caspase-3 activity [82].

Martín et al. [83] studied the potential protective effect of a polyphenolic extract from cocoa on cell viability and the antioxidant defenses of cultured human HepG2 cells subjected to oxidative stress by t-butyl hydroperoxide. Pretreatment of cells with 0.05–50 μg/mL of cocoa polyphenol extract for 2 or 20 h completely prevented cell damage and enhanced the activity of antioxidant enzymes. Moreover, the lower levels of GSH caused by t-butyl hydroperoxide in HepG2 cells were partly recovered through pretreatment with cocoa-enriched extract, which also prevented in a dose-dependent fashion the increase in ROS induced by t-butyl hydroperoxide. Cocoa polyphenols thus seem to confer significant protection against oxidation in these cells.

Cocoa polyphenol extracts dose-dependently (10–100 μM) attenuated hydrogen peroxide-induced inhibition of gap-junction intercellular communication (GJIC) in rat liver epithelial (RLE) cells and inhibited the hydrogen peroxide-induced phosphorylation and internalization of connexin 43, which is a regulating protein of GJIC in RLE cells. It also inhibited hydrogen peroxide-induced accumulation of ROS and activation of ERK. However, cocoa polyphenol extract did not block hydrogen peroxide-induced phosphorylation of MAPK. Similar results were observed ex vivo in RLE cell lysates in which cocoa polyphenol extract inhibited hydrogen peroxide-induced MAPK/

MEK1 activity. In these assays cocoa polyphenol bound directly with MEK1 to inhibit its activity. Inhibition of GJIC is strongly involved in carcinogenesis, particularly the tumor promotion process. Abnormal GJIC is attributable to phosphorylation of connexin 43, which is a major protein constituting gap-junction channels. The activation of ERK and MAPK signaling pathways is strongly related to both the phosphorylation of connexin 43 and the inhibition of GJIC in RLE cells. The authors thus concluded that cocoa polyphenols protect against hydrogen peroxide-induced inhibition of GJIC through antioxidant activity and direct inhibition of MEK activity, which may contribute to their chemopreventive potential [84].

Antiproliferative Effects

Treatment of cells with 50 μg/mL of procyanidin-enriched extracts inhibited cell growth by 70%, blocking the cell cycle in the G_2/M phase. The extracts also caused a significant decrease in the activities of ornithine decarboxylase and S-adenosylmethionine decarboxylase, two key enzymes of polyamine biosynthesis, thus leading to a decrease in the intracellular pool of polyamines. This could prove to be an important target in the antiproliferative effects of cocoa polyphenols [85].

ANTIMUTAGENIC AND CHEMOPROTECTIVE EFFECTS

Ohno et al. [86] found that cocoa extracts have an antimutagenic effect on chemical mutagens (benzo[a]pyrene) that require metabolic activation by cytochrome P450 (CYP), specifically CYP1A, in vitro. Cocoa products may thus prevent initiation of cancer by inhibiting the metabolic activation of carcinogens by CYP1A. The authors used the Ames test to study the effects of various enriched extracts of whole cocoa products on the mutagenicity of benzo[a]pyrene in Salmonella typhimurium strain TA 98 and t-butyl hydroperoxide in S. typhimurium strain TA 102. Benzo[a]pyrene induces mutagenicity through metabolic activation, whereas t-butyl hydroperoxide induces it through generation of free radicals. While white chocolate did not modulate the number of revertant colonies produced by benzo[a] pyrene treatment, milk chocolate and cocoa powder extracts did. Surprisingly, none of the cocoa products tested affected the number of revertant colonies produced by t-butyl hydroperoxide. Nevertheless, at maximum concentration (13.25 mg cocoa powder/mL), the crude cocoa powder extract reduced ethoxyresorufin O-deethylase activity to 17.4%, suggesting that whole cocoa products inhibit CYP1A activity, which may prevent DNA damage by reducing the metabolic activation of carcinogens.

In another study, Oleaga et al. [87] evaluated the effect of polyphenol cocoa extract in breast cancer cells (MCF-7 and SKBR3). After incubation with 250 ng/μL of cocoa extract, 7 genes out of 84 were overexpressed and 1 was underexpressed in MCF-7 cells, whereas 9 genes were overexpressed in SKBR3 cells. Because CYP1A1 was expressed in both cell lines, it was selected for further study. CYP1A1 mRNA, protein levels, and enzymatic activity all increased upon incubation with polyphenol cocoa extract. The combination of polyphenol cocoa extract with tamoxifen caused a synergistic cytotoxicity in both cell lines due to increased apoptosis of MCF-7 cells. The authors concluded that the interaction between estrogen receptor (ER)-α and aryl hydrocarbon receptor (AhR) upon incubation with polyphenol cocoa extract leads to the induction of CYP1A1 in breast cancer cells. The synergy observed between polyphenol cocoa extract and noncytotoxic tamoxifen concentrations opens the possibility for a combination therapy in which polyphenols from cocoa are used to boost the efficacy of tamoxifen.

A cocoa procyanidin fraction and procyanidin B2 at 5 μg/mL and 40 μM, respectively, inhibited TPA-induced neoplastic cell transformation of (JB6 P^+) mouse epidermal (JB6 P^+) cells by 47 and 93%, respectively, through the suppression of the kinase activity of MEK. They also inhibited the expression of COX-2, the TPA-induced activation of activator protein (AP)-1 and NF-κB, and the TPA-induced phosphorylation of MEK, ERK, and p90 ribosomal s6 kinase, all of which are implicated in inflammation and cancer [88].

Kim et al. [89] studied a cocoa polyphenol extract and its effect on vascular endothelial growth factor (VEGF) expression JB6 P^+ cells as well as its biological consequences, including its potential effects against proinflammatory cytokine-mediated skin cancer and inflammation. The cocoa polyphenol extract (5–20 μg/mL) inhibited TNF-α-induced upregulation of VEGF by reducing TNF-α-induced activation of the nuclear transcription factors AP-1 and NF-κB, which are key regulators of VEGF expression. The extract also inhibited TNF-α-induced phosphorylation of Akt and ERK, blocked activation of their downstream kinases (p70 ribosomal protein S6 kinase and p90 ribosomal protein S6 kinase), and suppressed phosphoinositide 3-kinase (PI3 K) activity by binding PI3 K directly. Moreover, although it did not affect the TNF-α-induced phosphorylation of MEK1, it suppressed TNF-α-induced MEK1 activity. The direct inhibition of PI3 K and MEK1 activities though the reduction of TNF-α-induced upregulation of VEGF may contribute to the chemopreventive potential of cocoa polyphenol. Indeed, the expression of VEGF, a multifunctional cytokine that is a key regulator of angiogenesis, is not only associated with tumorigenesis, but also with cardiovascular disease, rheumatoid arthritis, and psoriasis. Because cancerous cells require blood vessels to increase the flow of nutrients and oxygen vital for their development and survival via angiogenesis, the inhibition of VEGF expression may be a promising approach for protecting against carcinogenesis. Cocoa polyphenols could thus be considered potential chemopreventive agents against cancer and inflammation.

PHENOLS FROM COCOA IN TOOTH HEALTH

The production of caries and consumption of chocolate are often cited in the literature, with the intake of chocolate considered detrimental for tooth health. However, this negative effect is due to high concentrations of compounds present in chocolate other than cocoa, such as sugar. Ferrazzano et al. [90] were of the first to note the possible protective effect of cocoa on dental caries. Cocoa products contain inhibitors of the dextransucrase enzyme, which is responsible for the formation of the plaque extracellular polysaccharides from sucrose. Moreover, phenolic substances may be responsible for the observed anticaries effect of cocoa powder. Indeed, a water-soluble extract, of cocoa powder was shown to significantly reduce caries scores in rats infected withStreptococcus sobrinus, a potent cariogenic α-haemolytic streptococcus. Complementary studies have demonstrated that cocoa

consumption throughout the study period; the benefits observed with the cocoa extract are thus not due to dietary restriction. While the results of this study suggest that Acticoa powder may be beneficial in retarding age-related brain impairments, further studies are needed to elucidate the mechanisms of cocoa polyphenols in neuroprotection and to explore their effects in man [95]. Moreover, cocoa consumption protects nerves from injury and inflammation [93].

Antimalarial Effects

Numerous anecdotal reports of reduced episodic malaria in people from Ghana who drink a natural, unsweetened cocoa beverage daily prompted a search for scientific mechanisms to account for cocoa's possible antimalarial effects. In a review of the data from the literature on this subject, Addai [96] established five possible antimalarial mechanisms for cocoa. Thus, the effects could be due to the increased availability of antioxidants in plasma, membrane effects in general and in the erythrocyte membrane in particular, increased plasma levels of NO, specific antimalarial activity of cocoa flavanoids and their derivatives, and enhancement of the immune system mediated by components of cocoa including cocoa polyphenols.

Effects on Skin Tone and Elasticity

Cocoa polyphenols have a positive effect on skin structure when applied for at least 5 days. Studying these effects ex vivo, Gasser et al. [97] observed that cocoa polyphenols exhibit a positive action on the parameters assessed, including glycosaminoglycans and collagens I, III, and IV, associated with skin tone and elasticity. The doses at which they improve each of these parameters were determined to be 0.75% and 0.50%.

Mukai et al. [98] studied the suppression by a cocoa polyphenol extract (100 mg/kg, orally) of aryl hydrocarbon receptor transformation in C57BL/6 mice. Three hours after administration of the aryl hydrocarbon receptor agonist 3-methylcholanthrene (10 mg/kg, i.p.), the cocoa extract suppressed the induced transformation to control levels by inhibiting the formation of a heterodimer between the aryl hydrocarbon receptor and an aryl hydrocarbon receptor nuclear translocator in the liver. The extract also suppressed 3-methylcholanthrene-induced CYP1A1 expression and NAD(P)H: quinone-oxidoreductase activity while increasing glutathione S-transferase activity after 25 h. The authors concluded that the intake of cocoa polyphenol extract suppresses the toxicological effects of dioxins in the body [98]. Moreover, in topical preparations, protect the skin from oxidative damage caused by ultraviolet (UV) radiation [93].

CONCLUSIONS

Dark chocolate and cocoa enriched-polyphenol extracts may be beneficial for human health in different ways. Cocoa polyphenols have antioxidant properties, which endow them with various positive effects against several pathological disorders, including cardiovascular disease, inflammatory processes, and cancer. Indeed, cocoa polyphenols induce coronary vasodilatation, increase endothelial NO concentrations to induce vascular relaxation, improve vascular function, and decrease platelet adhesion. Moreover, they decrease levels of LDL-cholesterol and its oxidation while increasing HDL-cholesterol. Polyphenols also have anti-inflammatory activity, especially against IBD, through the inhibition of different transcription factor and cytokines. These effects also produce chemopreventive effects on other chronic diseases such as cancer by inhibiting the growth of various cancer cell lines. Of special interest are the effects of polyphenols on colon cancer. Indeed, the protective effects of polyphenols against IBD prevent it from evolving into cancer. In addition, it has been demonstrated that polyphenols cause nonapoptotic cell death, blocking the cell cycle in the G_2/M phase, which potentiates their antiproliferative effects. To summarize these effects, Table 1 lists the most relevant studies on cocoa phenolics and their findings, including the principal effects on metabolism, cardiovascular diseases, inflammation, and cancer prevention.

Table 1: Effects of cocoa phenolics on metabolism, cardiovascular diseases, inflammation, and cancer prevention

(a)

Metabolic and endocrine disorders		
	Assays	**Effects**
In vitro	Pancreatic α-amylase, pancreatic lipase, and secreted PLA_2 inhibitions [45]	(i) Cocoa extracts and procyanidins dose-dependently inhibited pancreatic α-amylase, pancreatic lipase, and secreted PLA_2, showing greater inhibitory activity against PLA_2. The inhibitory potency increases with the degree of polymerization
	HepG2 and Caco2 [50]	(i) (+)-catechin, (−)-epicatechin, procyanidin B2, procyanidin C1, and cinnamtannin A2 (10 μM): ApoA1 ¯ ApoB due to upregulation of SREBPs and increased LDL receptor activity
In vivo	Randomized crossover trial with 19 hypertensive patients with impaired glucose tolerance [47]	(i) Flavanol-rich dark chocolate (100 g/day for 15 days): (a) ¯ insulin resistance, systolic and diastolic blood pressure, total cholesterol, and LDL-cholesterol (b) insulin sensitivity, β-cell function, and flow-mediated dilation
	Randomized, crossover feeding trial in 42 high-risk volunteers [51]	(i) Chronic cocoa consumption (42 g/day for 4 weeks): (a) significant increases of phase II metabolites (glucuronide, sulfate conjugates) in 24 h urine (b) HDL-cholesterol (c) ¯ LDL-cholesterol
	Meta-analysis of 8 randomized controlled trials involving 215 participants [52]	(i) Short-term cocoa consumption: (a) ¯ LDL and total cholesterol (changes depended on the amount of cocoa consumed and the health status of participants)
	Randomized, single-blind, crossover study with 14 overweight/obese subjects [53]	(i) 20 g of dark chocolate with 500 mg of polyphenols and then 20 g of dark chocolate with 1000 mg of polyphenols or vice versa for 2 weeks separated by a 1-week washout period: (a) both doses were equally effective in reducing fasting blood glucose levels, systolic blood pressure, and diastolic blood pressure
	Randomized, placebo-controlled double-blind crossover trial [31]	(i) High polyphenol chocolate (16.6 mg of epicatechins in 45 g): (a) HDL-cholesterol (b) ¯ LDL-cholesterol, total cholesterol/HDL ratio High-polyphenol chocolate is effective in improving the atherosclerotic cholesterol profile in patients with diabetes

(b)	
Cardiovascular diseases	
Assays	**Effects**
Forty-two high-risk patients in a randomized crossover feeding trial for 4 weeks [32]	(i) 40 g of cocoa powder with 500 mL skim milk/day: (a) ¯VLA-4, CD40, and CD36 in monocytes (b) ¯ serum concentrations of P-selectin and ICAM-1
Resting human PBMCs from 13 healthy subjects treated with 25 μg/mL of procyanidin fractions isolated from cocoa [61]	(i) Individuals with low baseline levels of TGF-: TGF- release was enhanced in the range of 16–66%. Low-molecular-weight fractions (≤ pentamers) were more effective than their larger counterparts (hexamer or higher), with the monomer and dimer inducing the greatest increases (66% and 68%, resp.) (ii) Individuals with high baseline levels of TGF-: TGF- secretion was inhibited, being the inhibition most pronounced for trimmers through decamers (28–42%, resp.) and moderate for monomers to dimers (17–23%, resp.)
Twenty healthy subjects followed a balanced diet for 4 weeks; since day 14 to 27, they introduced daily 45 g of dark chocolate (860 mg of polyphenols, of which 58 mg were epicatechin) or white chocolate (5 mg polyphenols, undetectable epicatechin) [44]	(i) 2 h after dark chocolate intake (860 mg of polyphenols, of which 58 mg were epicatechin): (a) detectable epicatechin levels were observed (b) less DNA damage to mononuclear blood cells (c) no effect on plasma total antioxidant activity (ii) Effects were no longer evident after 22 h: dark chocolate
Blinded parallel-design study with 32 healthy subjects consuming 234 mg cocoa phenolics a day for 28 days [63]	(i) Plasma epicatechin and catechin concentrations by 81% and 28%, respectively (ii) ¯ Platelet function
Double-blind, randomized study with 22 heart transplant recipients [64]	(i) 2 h after ingestion of 40 g of flavonoid-rich dark chocolate (0.27 mg/g of catechin and 0.9 mg/g of epicatechin, with a total polyphenol content of 15.6 mg of epicatechin equivalents per gram): (a) coronary artery diameter (b) improved endothelium-dependent coronary vasomotion (c) ¯ platelet adhesion

Crossover, single blind study with 20 healthy subjects and 20 smokers who received 40 g of dark (cocoa > 85%) or milk (cocoa < 35%) chocolate [65]	(i) Smokers: (a) platelet recruitment, platelet formation of ROS and eicosanoids, and NOX2 activation (ii) Smokers + dark chocolate group: (a) ¯ platelet ROS, , and NOX2 activation decreased significantly (iii) Healthy + dark chocolate: (a) platelet variables did not change (iv) Milk chocolate (smokers and healthy): (a) no changes detected in either of the groups treated with milk chocolate
Single oral administration of a natural flavonoid-enriched cocoa powder (50–600 mg/kg) in spontaneously hypertensive rats [67]	Antihypertensive effect in hypertensive rats without modifying the arterial blood pressure in normotensive rats. No dose-response effect was observed
Fifty male Kurosawa and Kusanagi-hypercholesterolemic rabbits received 100 g/day of standard diet or cacao liquor polyphenol diet [68]	(i) Polyphenol-treated group: (a) ¯ area of atherosclerotic lesions in the aortas of the polyphenol-treated group was significantly smaller than in the control group (b) preserved parasympathetic nervous tone (c) no differences in the plasma lipid concentrations
Six-month clinical trial with 36 pre-hypertensive healthy adult volunteers, at content levels of 120–139 and 80–89 mmHg [70]	(i) 50 g of dark chocolate/day: (a) no significant differences were observed in the blood pressures of the treated and control groups
In vitro measurement of the kinetics of inhibition of purified angiotensin I converting enzyme from rabbit lung [72]	(i) Significantly inhibited the angiotensin I converting enzyme activity (ii) Monomeric flavan-3-ols: IC_{50} in the mM range (iii) Dimer and trimer: IC_{50} in the 100 μM range (iv) Larger procyanidins: IC_{50} in the 10 μM range

(c)		
Anti-inflammatory properties		
In vitro assays	**Polyphenol**	**Effect**
Unstimulated and LPS-stimulated PBMCs [20]	Different flavanol fractions (20 μg/mL)	(i) Monomers-pentamers: of LPS-induced synthesis of IL-1β, IL-6, IL-10, and TNF-α (ii) Long- and short-chain flavanol fractions: the production of GM-CSF in the absence of LPS and the expression of CD69 and CD83
Macrophages [73]	(i) Epicatechin (58 μg/mL) (ii) Cocoa extract (50 μg/mL) (iii) Isoquercitrin (18.6 μg/mL)	(i) ↓ MCP-1 and TNF-α at the transcriptional level (ii) ↓ TNF-α, IL-1α, and IL-6 mRNA levels (iii) ↓ NO secretion
THP-1 [72]	Procyanidin dimer B2: pretreatment for 30 min with 50 μM	(i) ↓ Expression of COX-2 (ii) ↓ Activation of ERK, JNK, and p38 MAPK (iii) Suppresses NF-κB activation through stabilization of IκB proteins
Caco-2 IL-1β-stimulated [77]	Polyphenol extract of cocoa (50 μmol of gallic acid/L); 4 h treatment	(i) In the absence of proinflammatory stimuli, the polyphenol extract of cocoa induces a basal PGE_2 synthesis due to COX-1 induction (ii) After 24 or 48 h stimulation: it decreased PGE_2 synthesis 4.6- and 2.2-fold, respectively
In vivo/Ex vivo	Polyphenol	Effect
DSS-induced ulcerative colitis in Balb/C mice [27]	Cocoa polyphenol-enriched extract (500 mg/kg)	(i) In vivo: ↓ weight loss, improves normal stool consistency, visible blood in feces, ↓ neutrophil infiltration, ↓ NO production, COX-2 expression, ↓ STAT3 and STAT1α phosphorylation (ii) Ex vivo in peritoneal macrophages: ↓ IL-6, IL-1β, and TNF-α (iii) In vitro in LPS-stimulated RAW 264.7 cells: inhibits NF-κB
Human PBMCs from 18 healthy volunteers [78]	Cocoa powder (40 g) with either milk or water	(i) Cocoa + water: (a) ↓ NF-κB activation in leukocytes and adhesion molecules (b) ↓ sICAM-1 (c) E-selectin (ii) Cocoa + milk: (a) no change in NF-κB activation (b) ↓ sICAM-1 (iii) Milk: (a) NF-κB activation
MCF-7 and SKBR3 [87]	(i) After incubation with 250 ng/μL of cocoa extract, 7 genes out of 84 were overexpressed and 1 was underexpressed in MCF-7 cells, whereas 9 genes were overexpressed in SKBR3 cells (ii) CYP1A1 mRNA, protein levels, and enzymatic activity increased (iii) The combination of polyphenol cocoa extract + tamoxifen caused a synergistic cytotoxicity	

RLE cells in vitro and ex vivo [84]	(i) Cocoa polyphenol extracts dose-dependently (10–100 μM) attenuated in vitro hydrogen peroxide-induced (a) inhibition of GJIC (b) phosphorylation and internalization of connexin 43 (c) accumulation of ROS and activation of ERK (ii) Ex vivo in RLE cell lysates (a) inhibits hydrogen peroxide-induced MAPK/ MEK1 activity
In vivo	
Animals	**Effects**
Wistar Han rats (5 weeks old) pretreated for 2 weeks with a cocoa-rich diet and injected with azoxymethane once a week for 2 weeks [82]	(i) The cocoa-rich diet (1 g of polyphenol/kg of diet): (a) antiproliferative effects in azoxymethane-induced colon cancer: ↓ ERK, Akt, and cyclin D1 (b) proapoptotic effects: ↓ Bcl-X_L levels and levels of Bax and caspase-3 activity

(d)

Cancer prevention	
In vitro	
Cell line	**Effects**
Caco-2 cells [85]	(i) Treatment of cells with 50 μg/mL of procyanidin-enriched extracts: (a) inhibits cell growth by 70%, blocking the cell cycle in the G_2/M phase (b) ↓ activities of ornithine decarboxylase and S-adenosylmethionine decarboxylase, and therefore, the intracellular pool of polyamines
PC12 rat pheochromocytoma cells [81]	(i) Cocoa procyanidin fraction (1 and 5 μg/mL) and procyanidin B2 (1 and 5μM): (a) ↓cell death attenuating the hydrogen peroxide-induced fragmentation of the nucleus and DNA ↓PARP cleavage, increased Bcl-X_L and Bcl-2 expression, and also inhibited activation of caspase-3 by hydrogen peroxide while attenuating the phosphorylation of JNK and MAPK
HepG2 cells [83]	(i) Pretreatment of cells subjected to oxidative stress with 0.05–50 μg/mL of cocoa polyphenol extract for 2 or 20 h: (a) completely prevented cell damage and enhanced the activity of antioxidant enzymes (b) recovered levels of GSH (c) prevented in a dose-dependent fashion the increase in ROS

Salmonella typhimurium strain TA 98 and TA 102 [86]	(i) Treatment with benzo[a]pyrene (a) white chocolate did not modulate the number of revertant colonies produced by treatment (b) milk chocolate and cocoa powder extracts did (ii) Treatment with -butyl hydroperoxide (a) none of the cocoa products tested affected the number of revertant colonies (b) 13.25 mg cocoa powder/mL reduced ethoxyresorufin -deethylase activity to 17.4% suggesting that whole cocoa products inhibit CYP1A activity
JB6 P+ cells [88, 89]	(i) Cocoa procyanidin fraction (5 μg/mL) and procyanidin B2 (40 μM) inhibit (a) TPA-induced neoplastic cell transformation at 47 and 93%, respectively (b) phosphorylation of MEK, ERK, and p90 ribosomal s6 kinase (c) COX-2 expression (d) AP-1 and NF-κB activation, and the TPA induced (ii) Cocoa polyphenol extract (5–20 μg/mL) (a) inhibits TNF-α-induced upregulation of VEGF by reducing TNF-α-induced activation of AP-1 and NF-κB (b) inhibits TNF-α-induced phosphorylation of Akt and ERK (c) suppresses PI3K activity by binding PI3K directly (d) suppresses TNF-α-induced MEK1 activity
MCF-7 and SKBR3 [87]	(i) After incubation with 250 ng/μL of cocoa extract, 7 genes out of 84 were overexpressed and 1 was underexpressed in MCF-7 cells, whereas 9 genes were overexpressed in SKBR3 cells (ii) CYP1A1 mRNA, protein levels, and enzymatic activity increased (iii) The combination of polyphenol cocoa extract + tamoxifen caused a synergistic cytotoxicity
RLE cells in vitro and ex vivo [84]	(i) Cocoa polyphenol extracts dose-dependently (10–100 μM) attenuated in vitro hydrogen peroxide-induced (a) inhibition of GJIC (b) phosphorylation and internalization of connexin 43 (c) accumulation of ROS and activation of ERK (ii) Ex vivo in RLE cell lysates (a) inhibits hydrogen peroxide-induced MAPK/MEK1 activity
In vivo	
Animals	**Effects**
Wistar Han rats (5 weeks old) pretreated for 2 weeks with a cocoa-rich diet and injected with azoxymethane once a week for 2 weeks [82]	(i) The cocoa-rich diet (1 g of polyphenol/kg of diet): (a) antiproliferative effects in azoxymethane-induced colon cancer: ↓ ERK, Akt, and cyclin D1 (b) proapoptotic effects: ↓ Bcl-X_L levels and levels of Bax and caspase-3 activity

ABBREVIATIONS, SYMBOLS, AND CELL LINES

ABTS:	-Azino-bis(3-ethylbenzthiazoline-6-sulphonic acid)
Akt:	Protein kinase B
AhR:	Aryl hydrocarbon receptor
AP:	Activator protein
Apo:	Apolipoprotein
Caco-2:	Human epithelial colorectal adenocar-cinoma cells
CD:	Cluster of differentiation
CD36:	Thrombospondin receptor
CD40:	TNF receptor superfamily member 5
cGMP:	Cyclic guanosine monophosphate
COX:	Cyclooxygenase
CYP:	Cytochrome P450
DNA:	Deoxyribonucleic acid
DPPH:	2,2-Diphenyl-1-pycrylhydrazyl
DSS:	Dextran sulfate sodium
EDHF:	Endothelium-derived hyperpolarizing factor
eNOS:	Endothelial nitric oxide synthase
ER-α:	Estrogen receptor-α
ERK:	Extracellular signal-regulated protein kinase
FRAP:	Ferric reducing-antioxidant power
GJIC:	Gap-junction intercellular communication
GM-CSF:	Granulocyte macrophage colony-stimulating factor
HDL:	High-density lipoprotein
HepG2:	Human hepatocellular carcinoma
IBD:	Inflammatory bowel disease
IC_{50}:	Inhibitory concentration-50
ICAM:	Intercellular adhesion molecule
IFN-γ:	Interferon-γ
Ig:	Immunoglobulin
IκB:	Inhibitor of κB
IL:	Interleukin

i.p.:	Intraperitoneally
JNK:	Jun-terminal kinase
LDL:	Low-density lipoprotein
LOX:	Lipoxygenase
LPS:	Lipopolysaccharide
LT:	Leukotrienes
MAPK:	Mitogen-activated protein kinase
MCF-7:	Breast cancer cells (Michigan Cancer Foundation-7)
MCP:	Monocyte chemoattractant protein
MEK:	Mitogen-activated protein kinase kinase
MMP:	Matrix metalloproteinase
MPO:	Myeloperoxidase
MT1:	Membrane type-1
NADPH:	Nicotinamide adenine dinucleotide phosphate
NF-κB:	Nuclear factor-κB
NO:	Nitric oxide
NOX:	NADPH oxidase
PARP:	Poly(ADP-ribose)polymerase
PBMCs:	Human peripheral blood mononuclear cells
PC12:	Rat pheochromocytoma cells
PG:	Prostaglandin
PHA:	Phytohemagglutinin
PI3 K:	Phosphoinositide 3-kinase
PLA_2:	Phospholipase A_2
RNA:	Ribonucleic acid
ROS:	Reactive oxygen species
RLE:	Rat liver epithelial
sICAM:	Soluble intercellular adhesion molecule; CD54
SKBR3:	Human breast carcinoma cells
SREBP:	Sterol regulatory element binding protein
STAT:	Signal transducers and activators of transcription
sVCAM:	Soluble vascular cell adhesion molecule; CD106
TGF-β:	Transforming growth factor-
THP-1:	Human monocytic cells
TNF-α:	Tumor necrosis factor-α
TPA:	12-O-Tetradecanoylphorbol-13-acetate
UV:	Ultraviolet
VEGF:	Vascular endothelial growth factor

VLA-4:	Very late antigen-4
VSMC:	Vascular smooth muscle cell.

ACKNOWLEDGMENTS

The authors thank the Spanish Government (MICIIN and FEDER, Grant SAF2009-10059-C03-01) for financial support. I. Andújar is grateful for a fellowship from the Generalitat Valenciana (Grant BFPI/2008/040).

REFERENCES

1. K. Szkudelska, L. Nogowski, and T. Szkudelski, "Resveratrol, a naturally occurring diphenolic compound, affects lipogenesis, lipolysis and the antilipolytic action of insulin in isolated rat adipocytes," The Journal of Steroid Biochemistry and Molecular Biology, vol. 113, no. 1-2, pp. 17–24, 2009
2. J. Ahn, I. Cho, S. Kim, D. Kwon, and T. Ha, "Dietary resveratrol alters lipid metabolism-related gene expression of mice on an atherogenic diet," Journal of Hepatology, vol. 49, no. 6, pp. 1019–1028, 2008.View at Publisher
3. C. Li, A. Allen, J. Kwagh et al., "Green tea polyphenols modulate insulin secretion by inhibiting glutamate dehydrogenase," The Journal of Biological Chemistry, vol. 281, no. 15, pp. 10214–10221, 2006.
4. M. Bose, J. D. Lambert, J. Ju, K. R. Reuhl, S. A. Shapses, and C. S. Yang, "The major green tea polyphenol, (−)-epigallocatechin-3-gallate, inhibits obesity, metabolic syndrome, and fatty liver disease in high-fat-fed mice," The Journal of Nutrition, vol. 138, no. 9, pp. 1677–1683, 2008.
5. M. A. Potenza, F. L. Marasciulo, M. Tarquinio et al., "EGCG, a green tea polyphenol, improves endothelial function and insulin sensitivity, reduces blood pressure, and protects against myocardial I/R injury in SHR," American Journal of Physiology, vol. 292, no. 5, pp. E1378–E1387, 2007.

6. A. Ejaz, D. Wu, P. Kwan, and M. Meydani, "Curcumin inhibits adipogenesis in 3T3-L1 adipocytes and angiogenesis and obesity in C57/BL mice," The Journal of Nutrition, vol. 139, no. 5, pp. 919–925, 2009.View at Publisher · View at Google Scholar · View at Scopus
7. S. Egert, A. Bosy-Westphal, J. Seiberl et al., "Quercetin reduces systolic blood pressure and plasma oxidised low-density lipoprotein concentrations in overweight subjects with a high-cardiovascular disease risk phenotype: a double-blinded, placebo-controlled cross-over study," British Journal of Nutrition, vol. 102, no. 7, pp. 1065–1074, 2009. View at Publisher
8. E. K. Kim, K. B. Kwon, M. Y. Song et al., "Flavonoids protect against cytokine-induced pancreatic β-cell damage through suppression of nuclear factor κB activation," Pancreas, vol. 35, no. 4, pp. e1–e9, 2007.
9. P. A. Ruiz, A. Braune, G. Hölzlwimmer, L. Quintanilla-Fend, and D. Haller, "Quercetin inhibits TNF-induced NF-κB transcription factor recruitment to proinflammatory gene promoters in murine intestinal epithelial cells," The Journal of Nutrition, vol. 137, no. 5, pp. 1208–1215, 2007.
10. S. Martínez-Flórez, B. Gutiérrez-Fernández, S. Sánchez-Campos, J. González-Gallego, and M. J. Tuñón, "Quercetin attenuates nuclear factor-κB activation and nitric oxide production in interleukin-1β-activated rat hepatocytes," The Journal of Nutrition, vol. 135, no. 6, pp. 1359–1365, 2005.
11. L. Rivera, R. Morón, M. Sánchez, A. Zarzuelo, and M. Galisteo, "Quercetin ameliorates metabolic syndrome and improves the inflammatory status in obese Zucker rats," Obesity, vol. 16, no. 9, pp. 2081–2087, 2008
12. S. Egert, C. Boesch-Saadatmandi, S. Wolffram, G. Rimbach, and M. J. Müller, "Serum lipid and blood pressure responses to quercetin vary in overweight patients by apolipoprotein E genotype," The Journal of Nutrition, vol. 140, no. 2, pp. 278–284, 2010.
13. E. P. Cherniack, "Polyphenols: planting the seeds of treatment for the metabolic syndrome," Nutrition, vol. 27, no. 6, pp. 617–623, 2011.

14. C. L. Keen, "Chocolate: food as medicine/medicine as food," Journal of the American College of Nutrition, vol. 20, supplement 5, pp. 436S–439S, 2001.

15. J. Wollgast and E. Anklam, "Polyphenols in chocolate: is there a contribution to human health?" Food Research International, vol. 33, no. 6, pp. 449–459, 2000.

16. G. Schinella, S. Mosca, E. Cienfuegos-Jovellanos et al., "Antioxidant properties of polyphenol-rich cocoa products industrially processed," Food Research International, vol. 43, no. 6, pp. 1614–1623, 2010. View at Publisher T. Hatano, H. Miyatake, M. Natsume et al., "Proanthocyanidin glycosides and related polyphenols from cacao liquor and their antioxidant effects," Phytochemistry, vol. 59, no. 7, pp. 749–758, 2002

17. J. H. Weisburger, "Chemopreventive effects of cocoa polyphenols on chronic diseases," Experimental Biology and Medicine, vol. 226, no. 10, pp. 891–897, 2001.

18. F. M. Steinberg, M. M. Bearden, and C. L. Keen, "Cocoa and chocolate flavonoids: Implications for cardiovascular health," Journal of the American Dietetic Association, vol. 103, no. 2, pp. 215–223, 2003.

19. C. Sanbongi, N. Suzuki, and T. Sakane, "Polyphenols in chocolate, which have antioxidant activity, modulate immune functions in humans in vitro," Cellular Immunology, vol. 177, no. 2, pp. 129–136, 1997.

20. T. P. Kenny, C. L. Keen, H. H. Schmitz, and M. E. Gershwin, "Immune effects of cocoa procyanidin oligomers on peripheral blood mononuclear cells," Experimental Biology and Medicine, vol. 232, no. 2, pp. 293–300, 2007.

21. E. Ramiro, A. Franch, C. Castellote, C. Andrés-Lacueva, M. Izquierdo-Pulido, and M. Castell, "Effect of Theobroma cacao flavonoids on immune activation of a lymphoid cell line," British Journal of Nutrition, vol. 93, no. 6, pp. 859–866, 2005.

22. E. Ramiro-Puig and M. Castell, "Cocoa: antioxidant and immunomodulator," British Journal of Nutrition, vol. 101, no. 7, pp. 931–940, 2009. View at Publisher

23. T. Pérez-Berezo, A. Franch, S. Ramos-Romero, C. Castellote, F. J. Pérez-Cano, and M. Castell, "Cocoa-enriched diets modulate intestinal and systemic humoral immune response in young adult rats,"Molecular Nutrition and Food Research, vol. 55, supplement 1, pp. S56–S66, 2011.
24. C. Selmi, T. K. Mao, C. L. Keen, H. H. Schmitz, and M. E. Gershwin, "The anti-inflammatory properties of cocoa flavanols," Journal of Cardiovascular Pharmacology, vol. 47, no. 2, pp. S163–S171, 2006.
25. G. W. Dryden, M. Song, and C. McClain, "Polyphenols and gastrointestinal diseases," Current Opinion in Gastroenterology, vol. 22, no. 2, pp. 165–170, 2006.
26. I. Andújar, M. C. Recio, R. M. Giner et al., "Inhibition of ulcerative colitis in mice after oral administration of a polyphenol-enriched cocoa extract is mediated by the inhibition of STAT1 and STAT3 phosphorylation in colon cells," Journal of Agricultural and Food Chemistry, vol. 59, no. 12, pp. 6474–6483, 2011.
27. W. Ren, Z. Qiao, H. Wang, L. Zhu, and L. Zhang, "Flavonoids: promising anticancer agents," Medicinal Research Reviews, vol. 23, no. 4, pp. 519–534, 2003
28. M. Rusconi and A. Conti, "Theobroma cacao L., the food of the Gods: a scientific approach beyond myths and claims," Pharmacological Research, vol. 61, no. 1, pp. 5–13, 2010.
29. F. A. Tomás-Barberán, E. Cienfuegos-Jovellanos, A. Marín et al., "A new process to develop a cocoa powder with higher flavonoid monomer content and enhanced bioavailability in healthy humans,"Journal of Agricultural and Food Chemistry, vol. 55, no. 10, pp. 3926–3935, 2007.
30. D. D. Mellor, T. Sathyapalan, E. S. Kilpatrick, S. Beckett, and S. L. Atkin, "High-cocoa polyphenol-rich chocolate improves HDL cholesterol in type 2 diabetes patients," Diabetic Medicine, vol. 27, no. 11, pp. 1318–1321, 2010.
31. M. Monagas, N. Khan, C. Andres-Lacueva et al., "Effect of cocoa powder on the modulation of inflammatory biomarkers in

patients at high risk of cardiovascular disease," American Journal of Clinical Nutrition, vol. 90, no. 5, pp. 1144–1150, 2009.

32. G. Williamson, "Bioavailability and health effects of cocoa polyphenols," Inflammopharmacology, vol. 17, no. 2, article 111, 2009.

33. R. R. Holt, S. A. Lazarus, M. C. Sullards et al., "Procyanidin dimer B2 [epicatechin-(4β-8)-epicatechin] in human plasma after the consumption of a flavanol-rich cocoa," American Journal of Clinical Nutrition, vol. 76, no. 4, pp. 798–804, 2002.

34. J. P. E. Spencer, F. Chaudry, A. S. Pannala, S. K. Srai, E. Debnam, and C. Rice-Evans, "Decomposition of cocoa procyanidins in the gastric milieu," Biochemical and Biophysical Research Communications, vol. 272, no. 1, pp. 236–241, 2000.

35. L. Y. Rios, R. N. Bennett, S. A. Lazarus, C. Rémésy, A. Scalbert, and G. Williamson, "Cocoa procyanidins are stable during gastric transit in humans," American Journal of Clinical Nutrition, vol. 76, no. 5, pp. 1106–1110, 2002.

36. M. P. Gonthier, J. L. Donovan, O. Texier, C. Felgines, C. Remesy, and A. Scalbert, "Metabolism of dietary procyanidins in rats," Free Radical Biology and Medicine, vol. 35, no. 8, pp. 837–844, 2003.

37. N. Ortega, J. Reguant, M. P. Romero, A. Macià, and M. J. Motilva, "Effect of fat content on the digestibility and bioaccessibility of cocoa polyphenol by an in vitro digestion model," Journal of Agricultural and Food Chemistry, vol. 57, no. 13, pp. 5743–5749, 2009.

38. C. Manach, A. Scalbert, C. Morand, C. Rémésy, and L. Jiménez, "Polyphenols: food sources and bioavailability," American Journal of Clinical Nutrition, vol. 79, no. 5, pp. 727–747, 2004.

39. K. W. Lee, Y. J. Kim, H. J. Lee, and C. Y. Lee, "Cocoa has more phenolic phytochemicals and a higher antioxidant capacity than teas and red wine," Journal of Agricultural and Food Chemistry, vol. 51, no. 25, pp. 7292–7295, 2003.

40. P. Schroeder, L. O. Klotz, D. P. Buchczyk, C. D. Sadik, T. Schewe, and H. Sies, "Epicatechin selectively prevents nitration but not oxidation reactions of peroxynitrite," Biochemical and Biophysical Research Communications, vol. 285, no. 3, pp. 782–787, 2001.

41. S. V. Verstraeten, J. F. Hammerstone, C. L. Keen, C. G. Fraga, and P. I. Oteiza, "Antioxidant and membrane effects of procyanidin dimers and trimers isolated from peanut and cocoa," Journal of Agricultural and Food Chemistry, vol. 53, no. 12, pp. 5041–5048, 2005.
42. P. Martorell, J. V. Forment, R. de Llanos et al., "Use of Saccharomyces cerevisiae and Caenorhabditis elegans as model organisms to study the effect of cocoa polyphenols in the resistance to oxidative stress,"Journal of Agricultural and Food Chemistry, vol. 59, no. 5, pp. 2077–2085, 2011.
43. A. Spadafranca, C. Martínez-Conesa, S. Sirini, and G. Testolin, "Effect of dark chocolate on plasma epicatechin levels, DNA resistance to oxidative stress and total antioxidant activity in healthy subjects,"British Journal of Nutrition, vol. 103, no. 7, pp. 1008–1014, 2010.
44. Y. Gu, W. J. Hurst, D. A. Stuart, and J. D. Lambert, "Inhibition of key digestive enzymes by cocoa extracts and procyanidins," Journal of Agricultural and Food Chemistry, vol. 59, no. 10, pp. 5305–5311, 2011.
45. K. Hanhineva, R. Törrönen, I. Bondia-Pons et al., "Impact of dietary polyphenols on carbohydrate metabolism," International Journal of Molecular Sciences, vol. 11, no. 4, pp. 1365–1402, 2010.
46. D. Grassi, G. Desideri, S. Necozione et al., "Blood pressure is reduced and insulin sensitivity increased in glucose-intolerant, hypertensive subjects after 15 days of consuming high-polyphenol dark chocolate,"The Journal of Nutrition, vol. 138, no. 9, pp. 1671–1676, 2008.
47. M. G. Shrime, S. R. Bauer, A. C. McDonald, N. H. Chowdhury, C. E. Coltart, and E. L. Ding, "Flavonoid-rich cocoa consumption affects multiple cardiovascular risk factors in a meta-analysis of short-term studies," The Journal of Nutrition, vol. 141, no. 11, pp. 1982–1988, 2011.
48. L. Hooper, C. Kay, A. Abdelhamid et al., "Effects of chocolate, cocoa, and flavan-3-ols on cardiovascular health: a systematic review and meta-analysis of randomized trials," American Journal of Clinical Nutrition, vol. 95, no. 3, pp. 740–751, 2012.

49. A. Yasuda, M. Natsume, N. Osakabe, K. Kawahata, and J. Koga, "Cacao polyphenols influence the regulation of apolipoprotein in HepG2 and Caco2 cells," Journal of Agricultural and Food Chemistry, vol. 59, no. 4, pp. 1470–1476, 2011.
50. N. Khan, M. Monagas, C. Andres-Lacueva et al., "Regular consumption of cocoa powder with milk increases HDL cholesterol and reduces oxidized LDL levels in subjects at high-risk of cardiovascular disease," Nutrition, Metabolism and Cardiovascular Diseases. In press.
51. L. Jia, X. Liu, Y. Y. Bai et al., "Short-term effect of cocoa product consumption on lipid profile: a meta-analysis of randomized controlled trials," American Journal of Clinical Nutrition, vol. 92, no. 1, pp. 218–225, 2010.
52. S. Almoosawi, L. Fyfe, C. Ho, and E. Al-Dujaili, "The effect of polyphenol-rich dark chocolate on fasting capillary whole blood glucose, total cholesterol, blood pressure and glucocorticoids in healthy overweight and obese subjects," British Journal of Nutrition, vol. 103, no. 6, pp. 842–850, 2010
53. R. Corti, A. J. Flammer, N. K. Hollenberg, and T. F. Lüscher, "Cocoa and cardiovascular health,"Circulation, vol. 119, no. 10, pp. 1433–1441, 2009.
54. M. Gómez-Juaristi, L. González-Torres, L. Bravo, M. P. Vaquero, S. Bastida, and F. J. Sánchez-Muniz, "Beneficial effects of chocolate on cardiovascular health," Nutricion Hospitalaria, vol. 26, no. 2, pp. 289–292, 2011.
55. O. Khawaja, J. M. Gaziano, and L. Djoussé, "Chocolate and coronary heart disease: a systematic review,"Current Atherosclerosis Reports, vol. 13, no. 6, pp. 447–452, 2011.
56. H. Sies, T. Schewe, C. Heiss, and M. Kelm, "Cocoa polyphenols and inflammatory mediators," The American Journal of Clinical Nutrition, vol. 81, no. 1, supplement, pp. 304S–312S, 2005.
57. V. B. Schini-Kerth, C. Auger, N. Étienne-Selloum, and T. Chataigneau, "Polyphenol-induced endothelium-dependent relaxations. Role of NO and EDHF," Advances in Pharmacology, vol. 60, pp. 136–175, 2010.
58. K. W. Lee, N. J. Kang, M. H. Oak et al., "Cocoa procyanidins inhibit expression and activation of MMP-2 in vascular smooth

muscle cells by direct inhibition of MEK and MT1-MMP activities," Cardiovascular Research, vol. 79, no. 1, pp. 34–41, 2008.

59. R. Corder, "Red wine, chocolate and vascular health: developing the evidence base," Heart, vol. 94, no. 7, pp. 821–823, 2008.

60. T. K. Mao, J. van de Water, C. L. Keen, H. H. Schmitz, and M. E. Gershwin, "Cocoa flavonols and procyanidins promote transforming growth factor-β_1 homeostasis in peripheral blood mononuclear cells," Experimental Biology and Medicine, vol. 228, no. 1, pp. 93–99, 2003.

61. E. L. Ding, S. M. Hutfless, X. Ding, and S. Girotra, "Chocolate and prevention of cardiovascular disease: a systematic review," Nutrition and Metabolism, vol. 3, article 2, 2006.

62. K. J. Murphy, A. K. Chronopoulos, I. Singh et al., "Dietary flavanols and procyanidin oligomers from cocoa (Theobroma cacao) inhibit platelet function," American Journal of Clinical Nutrition, vol. 77, no. 6, pp. 1466–1473, 2003.

63. A. J. Flammer, F. Hermann, I. Sudano et al., "Dark chocolate improves coronary vasomotion and reduces platelet reactivity," Circulation, vol. 116, no. 21, pp. 2376–2382, 2007.

64. R. Carnevale, L. Loffredo, P. Pignatelli et al., "Dark chocolate inhibits platelet isoprostanes via NOX2 down-regulation in smokers," Journal of Thrombosis and Haemostasis, vol. 10, no. 1, pp. 125–132, 2012.

65. M. Galleano, O. Pechanova, and C. G. Fraga, "Hypertension, nitric oxide, oxidants, and dietary plant polyphenols," Current Pharmaceutical Biotechnology, vol. 11, no. 8, pp. 837–848, 2010.

66. E. Cienfuegos-Jovellanos, M. Del Mar Quiñones, B. Muguerza, L. Moulay, M. Miguel, and A. Aleixandre, "Antihypertensive effect of a polyphenol-rich cocoa powder industrially processed to preserve the original flavonoids of the cocoa beans," Journal of Agricultural and Food Chemistry, vol. 57, no. 14, pp. 6156–6162, 2009.

67. M. Akita, M. Kuwahara, F. Itoh et al., "Effects of cacao liquor polyphenols on cardiovascular and autonomic nervous functions in hypercholesterolaemic rabbits," Basic and Clinical

Pharmacology and Toxicology, vol. 103, no. 6, pp. 581–587, 2008.

68. G. Rimbach, M. Melchin, J. Moehring, and A. E. Wagner, "Polyphenols from cocoa and vascular health—a critical review," International Journal of Molecular Sciences, vol. 10, no. 10, pp. 4290–4309, 2009.
69. K. Ried, O. R. Frank, and N. P. Stocks, "Dark chocolate or tomato extract for prehypertension: a randomised controlled trial," BMC Complementary and Alternative Medicine, vol. 9, article 22, 2009.
70. D. Grassi, G. Desideri, and C. Ferri, "Blood pressure and cardiovascular risk: what about cocoa and chocolate?" Archives of Biochemistry and Biophysics, vol. 501, no. 1, pp. 112–115, 2010.
71. L. Actis-Goretta, J. I. Ottaviani, and C. G. Fraga, "Inhibition of angiotensin converting enzyme activity by flavanol-rich foods," Journal of Agricultural and Food Chemistry, vol. 54, no. 1, pp. 229–234, 2006.
72. S. Desch, J. Schmidt, D. Kobler et al., "Effect of cocoa products on blood pressure: systematic review and meta-analysis," American Journal of Hypertension, vol. 23, no. 1, pp. 97–103, 2010.
73. E. Ramiro, A. Franch, C. Castellote et al., "Flavonoids from Theobroma cacao down-regulate inflammatory mediators," Journal of Agricultural and Food Chemistry, vol. 53, no. 22, pp. 8506–8511, 2005.
74. W. Y. Zhang, H. Q. Liu, K. Q. Xie et al., "Procyanidin dimer B2 [epicatechin-(4β-8)-epicatechin] suppresses the expression of cyclooxygenase-2 in endotoxin-treated monocytic cells," Biochemical and Biophysical Research Communications, vol. 345, no. 1, pp. 508–515, 2006.
75. B. Romier, Y. J. Schneider, Y. Larondelle, and A. During, "Dietary polyphenols can modulate the intestinal inflammatory response," Nutrition Reviews, vol. 67, no. 7, pp. 363–378, 2009.
76. B. Romier-Crouzet, J. van de Walle, A. During et al., "Inhibition of inflammatory mediators by polyphenolic plant extracts in human intestinal Caco-2 cells," Food and Chemical Toxicology, vol. 47, no. 6, pp. 1221–1230, 2009.

77. M. Vázquez-Agell, M. Urpi-Sarda, E. Sacanella et al., "Cocoa consumption reduces NF-κB activation in peripheral blood mononuclear cells in humans," Nutrition, Metabolism and Cardiovascular Diseases. In press.
78. H. Shapiro, P. Singer, Z. Halpern, and R. Bruck, "Polyphenols in the treatment of inflammatory bowel disease and acute pancreatitis," Gut, vol. 56, no. 3, pp. 426–436, 2007.
79. M. Yamagishi, N. Osakabe, M. Natsume et al., "Anticlastogenic activity of cacao: inhibitory effect of cacao liquor polyphenols against mitomycin C-induced DNA damage," Food and Chemical Toxicology, vol. 39, no. 12, pp. 1279–1283, 2001.
80. E. S. Cho, K. W. Lee, and H. J. Lee, "Cocoa procyanidins protect PC12 cells from hydrogen-peroxide-induced apoptosis by inhibiting activation of p38 MAPK and JNK," Mutation Research, vol. 640, no. 1-2, pp. 123–130, 2008.
81. I. Rodríguez-Ramiro, S. Ramos, E. López-Oliva et al., "Cocoa-rich diet prevents azoxymethane-induced colonic preneoplastic lesions in rats by restraining oxidative stress and cell proliferation and inducing apoptosis," Molecular Nutrition and Food Research, vol. 55, no. 12, pp. 1895–1899, 2011.
82. M. A. Martín, S. Ramos, R. Mateos et al., "Protection of human HepG2 cells against oxidative stress by cocoa phenolic extract," Journal of Agricultural and Food Chemistry, vol. 56, no. 17, pp. 7765–7772, 2008.
83. D. E. Lee, N. J. Kang, K. M. Lee et al., "Cocoa polyphenols attenuate hydrogen peroxide-induced inhibition of gap-junction intercellular communication by blocking phosphorylation of connexin 43 via the MEK/ERK signaling pathway," The Journal of Nutritional Biochemistry, vol. 21, no. 8, pp. 680–686, 2010.
84. S. Carnésecchi, Y. Schneider, S. A. Lazarus, D. Coehlo, F. Gossé, and F. Raul, "Flavanols and procyanidins of cocoa and chocolate inhibit growth and polyamine biosynthesis of human colonic cancer cells," Cancer Letters, vol. 175, no. 2, pp. 147–155, 2002.
85. M. Ohno, K. Q. Sakamoto, M. Ishizuka, and S. Fujita, "Crude cacao Theobroma cacao extract reduces mutagenicity induced by benzo[a]pyrene through inhibition of CYP1A activity in vitro," Phytotherapy Research, vol. 23, no. 8, pp. 1134–1139, 2009.

86. C. Oleaga, M. García, A. Solé, C. J. Ciudad, M. Izquierdo-Pulido, and V. Noé, "CYP1A1 is overexpressed upon incubation of breast cancer cells with a polyphenolic cocoa extract," European Journal of Nutrition, vol. 51, no. 4, pp. 465–476, 2012.

87. N. J. Kang, K. W. Lee, D. E. Lee et al., "Cocoa procyanidins suppress transformation by inhibiting mitogen-activated protein kinase kinase," The Journal of Biological Chemistry, vol. 283, no. 30, pp. 20664–20673, 2008.

88. J. E. Kim, J. E. Son, S. K. Jung et al., "Cocoa polyphenols suppress TNF-α-induced vascular endothelial growth factor expression by inhibiting phosphoinositide 3-kinase (PI3K) and mitogen-activated protein kinase kinase-1 (MEK1) activities in mouse epidermal cells," British Journal of Nutrition, vol. 104, no. 7, pp. 957–964, 2010

89. G. F. Ferrazzano, I. Amato, A. Ingenito, A. de Natale, and A. Pollio, "Anti-cariogenic effects of polyphenols from plant stimulant beverages (cocoa, coffee, tea)," Fitoterapia, vol. 80, no. 5, pp. 255–262, 2009.

90. T. Tomofuji, D. Ekuni, K. Irie et al., "Preventive effects of a cocoa-enriched diet on gingival oxidative stress in experimental periodontitis," Journal of Periodontology, vol. 80, no. 11, pp. 1799–1808, 2009.

91. T. K. Mao, J. van de Water, C. L. Keen, H. H. Schmitz, and M. E. Gershwin, "Effect of cocoa flavanols and their related oligomers on the secretion of interleukin-5 in peripheral blood mononuclear cells,"Journal of Medicinal Food, vol. 5, no. 1, pp. 17–22, 2002.

92. D. L. Katz, K. Doughty, and A. Ali, "Cocoa and chocolate in human health and disease," Antioxidant and Redox Signaling, vol. 15, no. 10, pp. 2779–2811, 2011.

93. T. Sathyapalan, S. Beckett, A. S. Rigby, D. D. Mellor, and S. L. Atkin, "High cocoa polyphenol rich chocolate may reduce the burden of the symptoms in chronic fatigue syndrome," Nutrition Journal, vol. 9, no. 1, article 55, 2010.

94. J. F. Bisson, A. Nejdi, P. Rozan, S. Hidalgo, R. Lalonde, and M. Messaoudi, "Effects of long-term administration of a cocoa polyphenolic extract (Acticoa powder) on cognitive performances in aged rats," British Journal of Nutrition, vol. 100, no. 1, pp. 94–101, 2008.

95. F. K. Addai, "Natural cocoa as diet-mediated antimalarial prophylaxis," Medical Hypotheses, vol. 74, no. 5, pp. 825–830, 2010.

96. P. Gasser, E. Lati, L. Peno-Mazzarino, D. Bouzoud, L. Allegaert, and H. Bernaert, "Cocoa polyphenols and their influence on parameters involved in ex vivo skin restructuring," International Journal of Cosmetic Science, vol. 30, no. 5, pp. 339–345, 2008.

97. R. Mukai, I. Fukuda, S. Nishiumi et al., "Cacao polyphenol extract suppresses transformation of an aryl hydrocarbon receptor in C57BL/6 mice," Journal of Agricultural and Food Chemistry, vol. 56, no. 21, pp. 10399–10405, 2008.

Chapter 15

Effect of Pre-Crystallization Process and Solid Particle Addition on Cocoa Butter Crystallization and Resulting Microstructure in Chocolate Model Systems

Lina Svanberg[a,b], Lilia Ahrné[a], Niklas Loréna, and Erich Windhab[b]

[a]SIK- the Swedish Institute for Food and Biotechnology, Box 5401, SE-402 29 Gothenburg, Sweden

[b]Swiss Federal Institute of Technology, Zurich, Institut für Lebensmittelwissenschaften, LFO E 12.1,ETH Zentrum, CH-8092 Zürich, Switzerland

ABSTRACT

The kinetics of cocoa butter crystallisation during solidification and resulting compactness of structure during storage for different chocolate model systems were investigated with respect to solid particle addition (sugar and cocoa particles) and pre-crystallization process (seeded/non-seeded). Confocal laser scanning microscopy (CLSM) was used to monitor microstructural evolution during solidification and image analysis were applied in order to quantify the kinetics. In order to quantify the compactness of structure during storage the migration rate of small-molecules was measured at different length scales. On the meso-scale, Fluorescence Recovery After Photobleaching (FRAP) was utilized to quantify local migration rate solely in the fat phase, whilst High Performance Liquid Chromatography (HPLC) measurements were performed to assess the global migration of same molecules on a macro scale. Both techniques were used in combination with microstructure characterization using CLSM and supported by differential scanning calorimeter melting curves for estimating cocoa butter polymorphism. During solidification, seeded samples tended to form multiple nucleation sites, inducing rapid growth of a crystal network. The non-seeded samples showed an altering structure, with some domains developing large spherical crystals while in other domains a more heterogeneous microstructure resulted. For the non-seeded samples, the impact of solid particles on the crystallization kinetics was also most pronounced. Both FRAP and HPLC analysis proved to generate relevant information of the effect of pre-crystallization and solid particles on compactness of structure during storage. FRAP- measurements gave detailed information of the hetero- or homogeneity in microstructure within the cocoa butter whilst the HPLC clearly showed the impact of solid particles. The combination of the two techniques revealed that a compact and homogeneous structure obtained through fast crystallization during solidification is required in order to retard global migration in confectionery systems.

INTRODUCTION

Fat bloom is a significant problem to the confectionary industry and is manifested as a greyish haze on the chocolate surface [1]. Filled confectionary products where the centre contains large amounts of highly mobile triacylglycerols (TAGs) are particularly sensitive to this phenomenon, as that filling fat migrates through the shell towards the surface where it can cause uncontrolled crystallisation and subsequently fat bloom. Additional quality defects associated with fat migration are shell softening, filling hardening, and sensory deterioration [2-4]. Chocolate has a complex microstructure in which sugar and cocoa particles are dispersed in a continuous phase of crystalline and liquid fat. Furthermore the crystalline cocoa butter can exist in six different polymorphic forms (often denoted by roman numbers I to VI and the Greek lettersa, b and b´). To retard migration, it is crucial that the cocoa butter is in a stable crystal form (b_V) together with a low fraction of liquid TAGs [5]. During chocolate manufacturing, the most common procedure for obtaining a stable b_V form involves subjecting the chocolate to a well-defined temperature programme under the action of shear. However, in 2000, Zeng presented a novel pre-crystallisation technique, "seeding technique", for producing well-tempered chocolate by homogeneously mixing 0.2-2% (w/w) of cocoa butter crystals in the most stable form ùVI with pre-cooled chocolate [6].

Confocal laser scanning microscopy (CLSM) offers the possibility to monitor microstructural development at different depths and length scales in the bulk under dynamic conditions which enables the survey of cocoa butter crystallisation. The technique can also be combined with fluorescence recovery after photobleaching (FRAP). Using the FRAP technique in combination with CLSM, it is possible to measure the mobility of fluorescent molecules directly in the microscope [7] and it has recently been used to correlate small-molecular diffusion through a cocoa butter matrix with the structural characteristics of the matrix [8]. In FRAP, first part of the fluorescent molecules within a well-defined area of the sample is deactivated using a high intensity laser beam that results in less intensity in the bleached area. Second, the fluorescent recovery, i.e., diffusion of unbleached molecules into the bleached area and simultaneous diffusion of bleached molecules into the surroundings, is detected as an increase in intensity in the bleached area [9, 10]. Parameters frequently derived from FRAP measurements are the mobile fraction and rate of mobility of the fluorescent molecules [11]. Consequently, FRAP offers the possibility to measure local diffusion of small florescence molecules within a structure. On a global scale, fat migration has previously been measured with high-performance liquid chromatography (HPLC) [12], flat bed scanner [13] or magnetic resonance imaging [14].

The objective of this work was to investigate the effect of improved/sub-optimal pre-crystallisation as well as impact of solid chocolate ingredients (sugar and cocoa particles) on the microstructure during cooling and storage in chocolate model systems. The CLSM technique was applied to follow fat crystallisation during cooling and the resulting microstructure was quantified using FRAP and HPLC during storage by measuring local and global diffusion properties of small molecules.

MATERIALS AND METHODS

Chocolate model systems were created by adding sugar or cocoa particles to cocoa butter according to the method described by [15]. To avoid particle density effects (as sugar is present in larger quantities than cocoa particles in commercial chocolate), the mass ratio between cocoa butter and solid particles was set to 2:1 with a total sample weight of 0.54 g. To visualise the crystallisation process with CLSM and enabling FRAP measurements the fluorescent labelled non-polar fatty acid analogue BODIPY FL C_{16} (Invitrogen Ltd., Paisley, UK) was mixed with the chocolate model systems while melted at a final concentration of 37 ppm. Subsequently, the fluorescent signal from BODIPY could be deactivated by a high intensity laser beam, thereby also enabling FRAP measurements. All model systems were subjected to one of two pre-crystallisation procedures: seeded or non-seeded, corresponding to improved and suboptimal processing followed by 60 minutes of cooling at 14°C according to the method described by [15].

The crystal growth in chocolate model systems during cooling at 14ÛC was followed using a Leica TCP SP2 confocal laser scanning microscope, (Mannheim, Germany). Images at a central point of each sample were taken with 1-2 minute interval using an HCX PL APO oil objective with 63 times magnification and a numerical aperature of 1.4. All images were recorded at a resolution of 1024×1024 pixels.

After 60 minutes of cooling at 14ÛC, FRAP measurements were made using a Leica TCP SP2 confocal laser scanning microscope (Mannheim, Germany) together with the rFRAP application available on the Leica software according to the procedure described by [16]. The size of the ROI was set to a 10 × 10 μm square, according to the rectangular FRAP (rFRAP) model developed by [17]. This relatively small ROI allowed measurement of the local diffusion rate of small molecules, i.e. BODIPY, solely in the fat crystal structure, so the ROI was always placed on the fat crystal networks visible in the frames. For each replicate in the model system, five rFRAP measurements were made immediately after 60 min of cooling and after one week of storage at 15°C. Only the rFRAP measurements where the experimental and model data coincided were included in the results. Calculations of the local diffusion rate of BODIPY, expressed in μm^2 /s, in the cocoa butter matrix were then made according to the methods of rectangular FRAP [17].

Global diffusion properties of the small molecule (BODIPY) were assessed with HPLC. The chocolate model systems used for HPLC analysis were either pure cocoa butter or cocoa butter mixed with both sugar and cocoa particles. Samples were also produced in which un-tempered cocoa butter (melted at 49°C for 10 min then cooled to 14°C for 60 min) and mixed with BODIPY to a final concentration of 100 ppm. Immediately after cooling, three samples representing 1) seeded/non-seeded cocoa butter, 2) untempered cocoa butter mixed with BODIPY, and 3) seeded/non-seeded cocoa butter, sugar, and cocoa particles, were placed into contact and held together by plastic foil. The samples were stored at 15°C for 5 d to allow BODIPY from the central layer to migrate into the adjacent model systems. After 5 d of storage, the three-layer systems were split to their original layers using a razorblade. Each separated model system was thereafter sliced into 8 thin layers (160 μm thick) using a Leica CM1900 cryostat (Mannheim, Germany). Every second slice was collected for HPLC analysis corresponding to original distances of 320, 640, 960, and 1280 μm from the cocoa butter/BODIPY layer. The amount of BODIPY in each slice was determined with HPLC according to the method described by [16]. All three-layer model systems were produced in duplicates.

RESULTS AND DISCUSSION

Microstructure during Cooling and Storage

CLSM micrographs illustrating the time-dependent structure evolution of seeded and non-seeded pure cocoa butter, cocoa butter and sugar and cocoa butter and cocoa particles are presented in Figure 1 and are adaptations of what has previously been published in [15, 16]. The staining dye (BODIPY) exhibited significantly higher affinity to the liquid cocoa butter compared to the crystals, thereby enabling the two phases to be distinguished by negative contrast. Both the sugar and cocoa particles appeared as black areas in the micrographs.

The structure formed during 60 minutes of cooling differed substantially between the two precrystallisation techniques. All seeded samples in Figure 1 formed multiple nucleation sites, which induced rapid growth of crystals within the first 15 minutes of cooling. The seeds were easily mixed in the samples, which resulted in a homogenous microstructure, with small proportions of liquid fat trapped within the fat crystal network. A more heterogeneous microstructure was observed in the non-seeded samples which coincide with previous reports, where the microstructure of poorly tempered chocolate has been investigated with various microscopy techniques [18-20]. The non-seeded samples containing cocoa particles formed small rod-shaped crystals (Figure 1(c)) which differed significantly from the homogenous microstructure formed in corresponding seeded samples where no such crystal morphology was detected. Further results can be observed in Svanberg et al., 2010 [15].

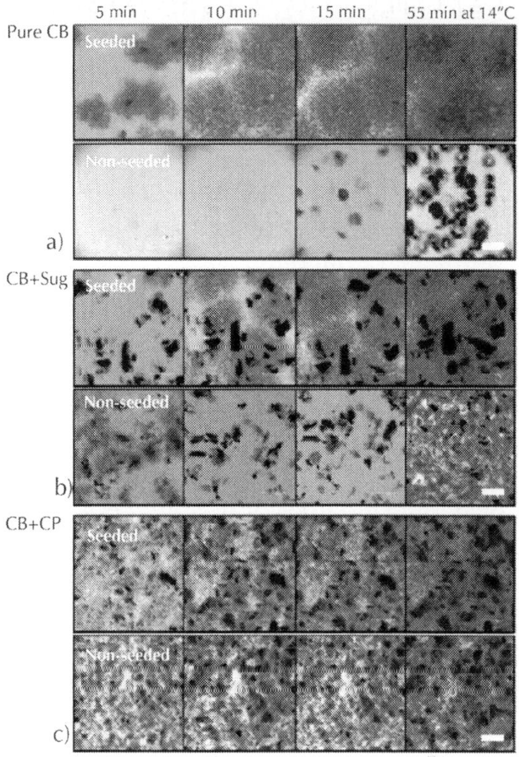

Figure 1: CLSM micrographs showing the time-dependant crystal growth in seeded and non-seeded (a) pure cococa butter (CB), (b) cocoa butter and sugar (CB+Sug) and (c) cocoa butter and cocoa particles (CB+CP), during cooling at 14ÛC for 55 minutes. Scale bars represent 25 μm.

During storage, extensive post-crystallisation occurred in the non-seeded samples and after one week the microstructure largely resembled the seeded samples. This is illustrated in Figure 2 where CLSM micrographs of the microstructure in seeded and non-seeded model systems after one week of storage are displayed.

Structure Density during Storage

Immediately after 60 min of cooling, where significant differences in microstructure were observed between the seeded and non-seeded samples, diffusion measurements with FRAP and HPLC were initiated to correlate the quantification of structure density with the results of the microstructure characterization. The results presented here are adaptations of what has been published in [16].

The calculated local diffusion rates of BODIPY, obtained from FRAP measurements, for all chocolate model systems after 60 min of cooling and after one week of storage are presented in Figure 3. Directly after cooling, the non-seeded samples had significantly higher local diffusion rates of BODIPY compared to the seeded samples. However, for the non-seeded sample with CB+sug the standard deviation was too large for any statistical conclusions. Initially, the substantial error bars in the non-seeded samples indicate larger variations in microstructure in the fat phase. This also agrees with previous findings regarding the compactness of structure density found in properly/improperly traditionally pre-crystallized chocolate[1, 2]. The seeded samples had markedly low local diffusion rates of BODIPY; however, it is important to note that all these samples are close to the detection limit for the applied FRAP method, which is 0.1μm^2/s [17]. After one week of storage, the differences between the seeded and non-seeded samples were less pronounced, as the local diffusion rate of BODIPY in the non-seeded samples had decreased significantly (Fig. 2(b)). The seeded samples displayed no such alteration during storage and the local diffusion rate remained constantly low. Furthermore, no significant differences in local diffusion rate were observed between chocolate model systems subjected to the same pre-crystallisation process i.e. possible effects of solid particles on structure density in the fat phase could not be significantly distinguished using the FRAP technique.

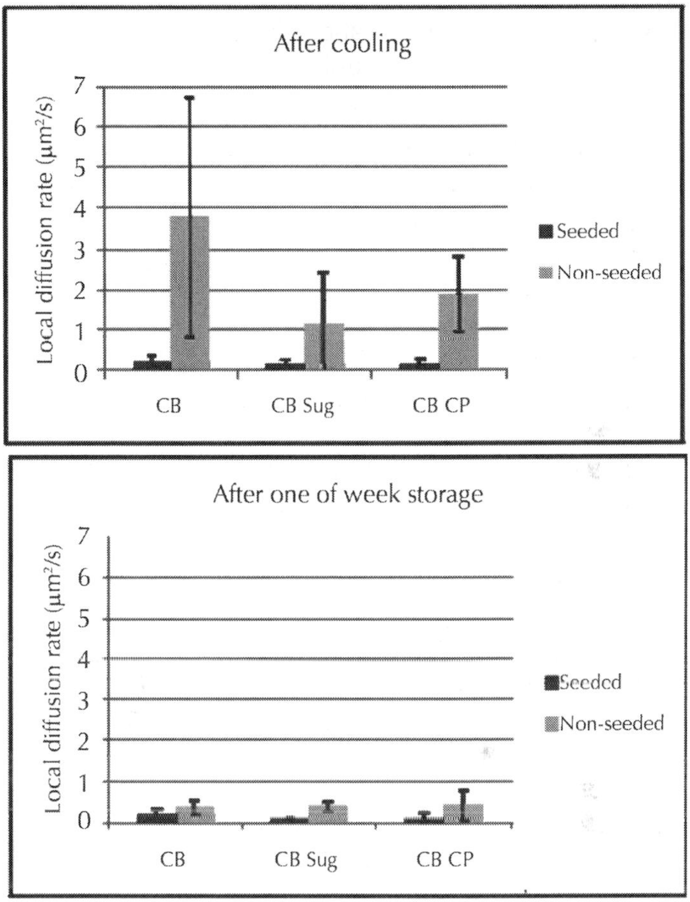

Figure 2: Calculated local diffusion rate of BODIPY (µm^2 /s) in seeded versus non-seeded chocolate model systems immediately after cooling (a) and after one week of storage at 15°C (b). The error bars represent the standard deviation of fifteen FRAP measurements divided on three sample measurements.

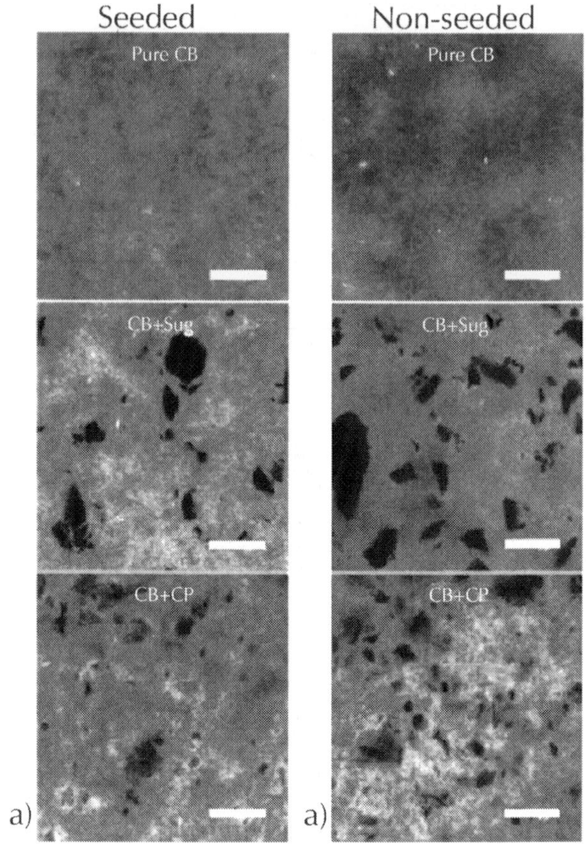

Figure 3: Representative CLSM micrographs of the microstructure in seeded (a) and non-seeded (b) chocolate model systems after one week of storage at 15°C. Scale bars represent 25 μm.

To further quantify the structure density in the various microstructures created by pre-crystallisation and/or solid particle addition, the global diffusion properties of BODIPY were determined using HPLC.

A high capacity to retard diffusion was treated as equivalent to a dense structure. The amount of BODIPY, expressed in μg of BODIPY/mg sample, as a function of distance from the cocoa butter/BODIPY slice is presented in Figure 4. A more detailed description of the structure density results can also be found in [16].

In the area nearest the cocoa butter/BODIPY slice, the amount of BODIPY was higher in the nonseeded samples for both pure cocoa butter and samples with solid particles (Figure 4(a)). This indicates that the microstructure in these samples had less ability to withstand the diffusion of BODIPY.

Furthermore, when comparing samples subjected to same pre-crystallisation, the addition of solid particles tended to retard the global diffusion. This could be due to an increase in tortuosity, i.e., deviation from the linear diffusion path as the solid particles forced the BODIPY to diffuse round them. This also agrees with previous reports examining the diffusion properties of various filling fats through a chocolate shell in the presence and absence of solid particles [2, 21, 22].

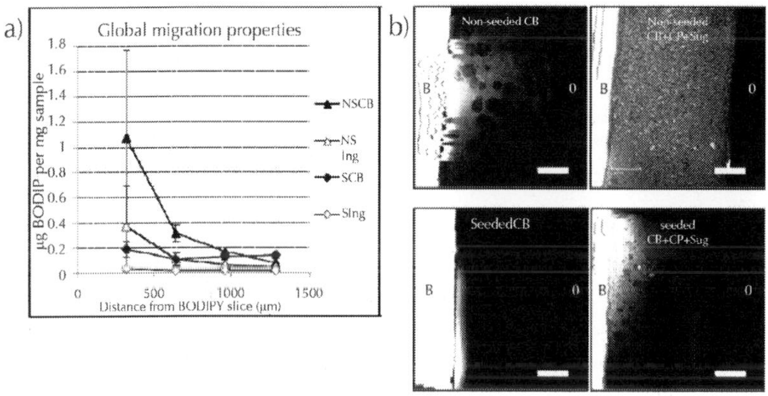

Figure 4: (a) Amount of BODIPY (μg of BODIPY/mg sample), measured with HPLC, as a function of distance from the cocoa butter/BODIPY slice. Error bars represent the standard deviation of two sample replicates. (b) Low-magnification CLSM micrographs of the global diffusion of BODIPY from the cocoa butter/BODIPY slice [B] into the model systems. The outer surface is indicated by [O] and scale bars represent 500 μm.

Low-magnification CLSM micrographs of the three-layer systems after 5 d of storage are presented in Figure 4(b). The bright area to the left in each micrograph represents to the cocoa butter/BODIPY slice, indicated by [B]. The investigated chocolate model systems correspond to the adjacent dark or orange area to the right, which is visible only due to diffusion of BODIPY. The outer surface, [O], was completely black in all the micrographs, and the solid particles appeared as dark spots in the CB+CP+Sug layer (Figure 3(b)). Since all the micrographs presented were taken using the same laser intensity, the significantly darker micrographs of the seeded samples correspond to less diffusion of BODIPY. Larger amounts of BODIPY had migrated through the non-seeded samples, which were manifested as bright/white areas in the micrographs.

From the microstructure characterization with CLSM it was found that post-crystallisation in the cocoa butter compensated for the initial heterogeneous microstructure found in the non-seeded samples.

However, the first days with a less dense structure, when crystallisation was not completely developed, were enough to allow significantly higher levels of diffusion to occur. It can be concluded that a dense structure is required immediately after cooling to retard diffusion and this is special relevant in filled chocolate products.

CONCLUSIONS

The CLSM was found to be a good technique for monitor the cocoa butter crystallisation in chocolate model systems during cooling. Seeded samples formed multiple nucleation sites, which induced a rapid growth of crystals and resulted in a more homogenous microstructure. The non-seeded samples showed a more random structure, with some areas developing large spherical crystals while other parts gained a more heterogeneous microstructure with large inclusions of liquid fat and small compact crystals.

The structure density in the various model systems during storage could be quantified by smallmolecule diffusion at different length scales using FRAP and HPLC. FRAP measurements gave detailed information on the heterogeneity or homogeneity of microstructure in the cocoa butter, whereas the HPLC indicated the impact of solid particles. Both techniques showed significant differences in terms of structure density between the pre-crystallisation processes, seeded and non-seeded. During storage, the non-seeded samples experienced extensive post-crystallisation and evolved towards the dense structure found in the seeded samples. However, the initial less compact structure in the non-seeded samples created a more favourable environment in which the global diffusion could occur.

REFERENCES

1. R.W. Hartel, 1999, Chocolate: Fat Bloom During Storage, The Manufacturing Confectioner. 79(5): p. 89-99.
2. V. Ghosh, G.R. Ziegler and R.C. Anantheswaran, 2002, Fat, Moisture and Ethanol Migration through Chocolates and Confectionary Coatings, Critical Reviews in Food Science and Nutrition. 42(6): p. 583-626.
3. G. Talbot, 1989, Fat migration in confectionery products, Confectionery Production. 55: p. 655-656.
4. G. Ziegleder, 1997, Fat migration and bloom, The Manufacturing Confectioner. 77(2): p. 43-44.
5. G.R. Ziegler, A. Shetty and R.C. Anantheswaran, 2004, Nut Oil Migration Through Chocolate, The Manufacturing Confectioner. 84(9): p. 118-126.
6. Y. Zeng, Impf- und Scherkristallisation von Schokoladen. PhD Thesis, in Laboratorium fur Lebensmittelverfahrenstechnik. 2000, ETH-Eidgenössische Technische Hochschule Zürich.
7. N. Lorén, M. Nydén and A.-M. Hermansson, 2009, Determination of local diffusion properties in heterogeneous biomaterials, Advances in Colliod and Interface Science. 150: p. 5-15.
8. S. Marty, M. Schroeder, K.W. Baker, G. Mazzanti and A.G. Marangoni, 2009, Small-Molecule Diffusion through Polycrystalline Triglyceride Networks Quantified Using Fluorescent Recovery after Photobleaching, Langmuir. 25(15): p. 8780-8785.
9. D. Axelrod, D.E. Koppel, J. Schlessinger and E. Elson, 1976, Mobility measurement by analysis of fluorescence photobleaching recovery kinetics, Biophysical Journal. 16: p. 1055-1069.
10. R. Peters, J. Peters, K.H. Tews and W. Bähr, 1974, A microfluorimetric study of translational diffusion in erythrocyte membranes, Biochimica et Biophysica Acta. Biomembranes. 367: p. 282-294.
11. E.A.J. Reits and J. Neefjes, 2001, From fixed to FRAP: measuring protein mobility and activity in living cells, Nature Cell Biology. 3: p. E145-E147.

12. R.S. Khan and D. Rousseau, 2006, Hazelnut oil migration in dark chocolate- kinetic, thermodymnamic and structural considerations, European Journal of Lipid Science and Technology. 108: p. 434-443.
13. S. Marty, K. Baker and A.G. Marangoni, 2009, Optimization of a scanner imaging technique to accurately study oil migration kinetics, Food Research International. 42(3): p. 368-373.
14. M.E. Miquel, S. Carli, P.J. Couzens, H.J. Wille and L.D. Hall, 2001, Kinetics of the migration of lipids in composite chocolate measured by magnetic resonance imaging, Food Research International. 34: p. 773-781.
15. L. Svanberg, L. Ahrné, N. Lorén and E.J. Windhab, 2010, Effect of Sugar, Cocoa Particles and Lecithin on Cocoa Butter Crystallisation in Seeded and Non-Seeded Chocolate Model Systems, Accepted for publication in Journal of Food Engineering.
16. L. Svanberg, L. Ahrné, N. Lorén and E.J. Windhab, 2011, Effect of pre-crystallization process and solid particle addition on microstructure in chocolate model systems, Accepted for publication in Food Research International.
17. H. Deschout, J. Hagman, S. Fransson, J. Jonasson, M. Rudemo, N. Lorén and K. Braeckmans, 2010, A Versatile Rectangle FRAP Method for Diffusion Measurements with a Laser Scanning Microscope, Biomacromolecules.
18. E.O. Afoakwa, A. Paterson, M. Fowler and J. Vieira, 2008, Effects of tempering and fat crystallization behaviour on microstructure, mechanical properties and appearance in dark chocolate systems., Journal of Food Engineering. 89: p. 128-136.
19. A. Bowser, 2006, Crystallization of cocoa butter, The Manufacturing Confectioner. 86(9): p. 115-118.
20. Y. Kinta and R.W. Hartel, 2010, Bloom Formation on Poorly-Tempered Chocolate and Effects of Seed Addition., Journal of the American Oil Chemists› Society. 87(1): p. 19-27.
21. W.L. Lee, M.J. McCarthy and K.L. McCarthy, 2010, Oil Migration in 2-Component Confectionery Systems, Journal of Food Science. 75(1): p. E83-E89.

22. D. Rousseau, The microstructure of chocolate, in Understanding and controlling the microstructure of complex foods, D.J. McClements, Editor. 2007, Woodhead Publishing Limited: Cambridge. p. 648-687.

Chapter 16

NGO as Triple-Helix Axis: Some Lessons from Nias Community Empowerment on Cocoa Production

Corinthias Pamatang Morgana Sianipara, b, Kitri Widaretnac

[a]Institute of Technology Bandung (ITB), Jl. Ganesha 10 (Gedung SBM-ITB), Bandung 40132, Indonesia

[b]Polytechnic for Manufacturing (POLMAN) Bandung, Jl. Kanayakan 21, Bandung 40135, Indonesia

[c]Swisscontact Indonesia, Gunungsitoli-Nias (NISA Project) & Maumere-Flores (LED-NTT & WISATA Projects), Indonesia

ABSTRACT

Community empowerment is one of the main purpose of development in developing countries. While development only result in a ratio between the "before" and "after" condition, empowerment can keep the continuity of the development through the good relationship between related parties, in terms of knowledge production and field application. Since the parties in community empowerment are tightly related with Triple-Helix model, Non-governmental organization (NGO), as a non-profit organization in community empowerment projects, should catch its opportunities as the axis of those three helixes. The case of Nias community empowerment on cocoa production presents some lessons which seek to illustrate the links between parties in a community empowerment project. While each helix make their movements by themself, the NGO fill-in the gap in the interactions between the helixs as well as interaction among each helix. The convergence between helixs build the opportunities for each connection, so it is analyzed carefully through internal informations to find the most appropriate relationship in the field. By focusing on the gathering of new thought of the Triple-Helix approach, this paper proposes a simple model as the initial step to develop new applications of the concept, then emerges a new thought that Triple-Helix can be implemented in wider phenomenon among societies, and give opportunities for further research around it.

INTRODUCTION

The dynamics of global economics in the last decades were noted through the change of world economic power. Many international agreements especially in economics issues were done more often, or with a bigger scope, between developed countries to developing countries. Months ago, Indonesia's return to investment-grade status from BB+ to BBB- (Fitch Ratings, 2011), the same status ever obtained 14 years ago. It means that Indonesia become a country with bigger opportunities for investors. In other hand, Indonesia is a developing country with natural resources as its base in economics. The structure of Indonesian export is still dominated by primary commodities, such as oil & gas and agricultural products (Adam & Negara, 2010).

Above facts show that Indonesia is an emerging country which has big opportunities to develop the investment environment through the development in primary commodities. The development will increase the bargaining position of Indonesia among the bilateral and multilateral agreements. Free-trade which is started in 2010 should give opportunities for each country among an agreement through specialization in superior commodity production, especially in hot commodities in the world like agricultural commodities (Hutabarat, *et al.*, 2007).

One of the Indonesia's hottest agricultural commodity to the world is cocoa. For the last 20 years Indonesia was the world's third largest cocoa producer after Ivory Coast and Ghana, contributing export earnings in excess of US$1.4 billion per year (Asia Pulse, 2011). In mid-2008, the Indonesian Government announced a large national program to revitalize cocoa industry, known as Gernas Pro Kakao. The key view is the 55% cocoa production growth to 2014/15 (Wood, 2010). Government initiatives to replace ageing trees will see an increase in yields as well as planted area, allowing Indonesia to compete with Ghana, the world's second largest producer.

LITERATURE REVIEW

Nias Cocoa Production

Cocoa (*Theobroma cacao L.*), one of the world's most important perennial crops, is primarily used for chocolate production, with an estimated global output of 3.5 million tons of beans being produced in 2006 (ICCO, 2007). The area of cacao plantations has increased greatly during the last decade in Indonesia, now the largest cocoa producer in South-East Asia, following the decline of Malaysia production in the last 20 years (Moser, *et al.*, 2010). Indonesia has 1.5 million hectares of cocoa plantations, mostly in the eastern Sulawesi island, with an extra 10,000 to 20,000 hectares of additional production due this year (Wall Street Journal, 2011).

Nias is the biggest island in Nias archipelago. On the edge of Indian ocean, the archipelago is included in North Sumatera province. The island is divided into four Kabupaten and one Kotamadya. Nias island is the 3rd biggest cocoa agricultural area in Indonesia, after Central Sulawesi and Flores (de Blij & Müller, 2010). The production output of cacao in Kabupaten Nias is the 3rd biggest commodity, after rubber and coconut (the datas are combined with Kotamadya Gunungsitoli, Kabupaten Nias Utara, and Kabupaten Nias Barat). Badan Pusat Statistik (BPS, 2009) have noted that in 2008, rubber commodity reach 1.170 tons productivity output, coconut in 964 tons level, and cacao with 830 tons production of dry bean.

The selling price is decided by the production volume and the quality of the output. Related to the increase of productivity, there should be an extensification and intensification of agricultural. In other hand, quality level should be improved through post-harvest treatment. They are processed as on-farm or upstream system. The intensification effort for cocoa commodity in Nias had been done by many local bodies or international institutions. Many of them were applied as joint-cooperations with local government in the recovery phase after tsunami (Aceh-Nias in 2004) and earthquake (Nias in 2005) disasters, for example in Kecamatan Bawolato, Idanögawo, and Gunungsitoli Idanoi (Swisscontact, 2011). In short, Nias cocoa farmers have had the basic knowledges of cocoa cultivation method. So, that is a good time to bring better development of post-harvest treatment in order to increase the quality of cocoa commodity. Usually farmers sell the dry bean. It is produced through post-harvest process. The commodity will be sold to the local buyers and exporters.

Harvest and post-harvest treatment are the most important things in cocoa cultivation, because both decide the output quality of cocoa commodity. Although the high production level, the shipping amount will decrease when the harvest and post-harvest process did not done well. Furthermore, bad quality dry cocoa bean means the buyer will not place the price appropriately, even; the exporters eventually deny the dry cocoa bean supply from the farmers. Because of that, the quality must be kept at –minimum– the same level with the trading standard through good post-harvest processing. Through the above facts, Swisscontact (hereinafter will be referred as only "the NGO") enter the field to develop the post-harvest system for Nias cocoa farmers. It is a non-governmental organization (NGO) from Switzerland, with tagline "We Create Opportunities." Founded in 1959 by the cooperation between private sector and academic sector in Switzerland, it is funded by private donors as well as Switzerland government (Swisscontact, 2010). Its activity in Nias is funded by Swiss Solidarity which based in Jenewa, Switzerland. They give the name for the project "NISA," the acronym from *Nias Income-generation through Sustainable Agriculture*. The project was started in January 2009.

Unlike the downstream processing steps in the northern hemisphere, the upstream processing steps rarely involve sophisticated equipment (Ng, 2011). In post-harvest cocoa processing, the quality of dry cocoa bean is tightly related with fermentation and drying process. The main idea in drying process is to decrease the water percentage in cocoa been to meet the trading standard requirement (Ülrich, 2009). One thing that highly influences the drying process performance is the local environment, so the drying technique must give more attention at the local condition (Sianipar, 2010b). Furthermore, it is stated that the most appropriate technique of cocoa drying in Nias was solar-powered drying, or using the sun light as the source of heat. Amin (2009) explained six main principles in drying cocoa bean using sun light:

- ***Drying Base***: The good drying base is using conductor material to absorb more heat from the sun light. More light means more heat, so the drying process will be done faster. Farmers usually use thick plastic sheet, concrete floor, or just spread the bean on the asphalt street.
- ***Rain-Proof***: Mobile roofs or mobile floors can be used to keep safe the cocoa bean from rain.
- ***Color & Material***: Dark colors and steel-type material are often used around the drying area to collect more heat from the sun.
- ***Heat Collector***: The shape and position of the heat collector must produce high temperature to get faster evaporation of water from the cocoa bean.
- ***Airflow***: The airflow must be guided to attach more water but does not reach saturation level. The air should keep in fast flow but must reach the whole skin of cocoa bean.
- ***Heat Isolator***: The heat must be kept in some level of temperature. The cocoa bean should be kept in hot or warm condition to prevent the absorption of water in the night or rainy day.

On the field, smallholder farmers are responsible for roughly seventy percent of total global cocoa production (Clay, 2004; Donald, 2004). An estimated 5–6 million smallholder farmers earn most or all of their cash income from cocoa production (Clay, 2004). Researcher as the living-pillar of academic sector must catch the opportunities to help them. Academicians should apply their knowledge to the society. Capitalization of knowledge happens when knowledge generates an economic added value (Viale, 2010). It means that researchers should use their knowledge to give an economic added value for the farmers through their innovations. Furthermore, the increase of economic value of the farmers will also increase the business opportunities for them.

The higher quality means the higher selling price of the dry cocoa bean. Everything which related to the optimization of drying process should also be capitalized to bring the empowerment of society.

Triple-Helix

The capitalization of knowledge should also be attributed as the result of intellectual property of the researcher. As one of the three-pillars of Indonesian education, social responsibility from the academic sector to the society should be mediated by the government. Dalle (2003) and Maurer (2002) have documented that many of research products are distributed and transferred to others using institutions that range from commercial. These institutions should be the government representations as the neutral side, from the appropriate ministry or department. Government can behave as the policy-maker that accommodate the intellectual rights of researcher, or behave as the neutral funding committee in the research as well as the application of the research result among the society. There should be a framework that highlights the difficulties in sustaining the production of knowledge when it is the outcome of a collective enterprise (Gambardella & Hall, 2010).

The connection between researchers as university representative, farmers as business side, and government as the policy-maker as well as the funding committee build the scheme of Triple-Helix. The triple helix is a model for capitalizing knowledge in order to pursue innovation (Etzkowitz, 2008). The triple-helix model seems to play an anti-cyclic role in innovation (Viale, 2010). It expresses its potential when the interaction between researcher, farmers, and government is not autonomous, but in dynamic way of the society. It works as a "nudge tool" (Thaler & Sunstein, 2008), whose aim is to maintain a sufficient flow of innovation through the right incentives and institutional mechanisms for academy –cocoa farming industry collaboration.

Theoretical Framework

The fails and success of the sustainable economic development is tightly influenced by the way the government do the governance (Firdausy, 2010). The statistics of economic development often seen from the contribution of industrial sector output to the economics of the country or the related area. In short, the output level is decided by the combination of economics inputs, such as manpower, financial investment, and technology (Nugroho & Hidayat, 2010). Because of that, the three helix in Nias case should be understood as business, academic, and government which are collaborating into the field to form a triple-helix, the three are interrelated in terms of investment, labor, and technology. Non-governmental organization (NGO), as a nonprofit organization, should catch it's opportunities as the axis of coordination for the triple-helix model. The discussion should find the position of NGO in filling the gap between those three helix to incorporate three economics inputs as the foundation to reach Nias sustainable economic development on cocoa production.

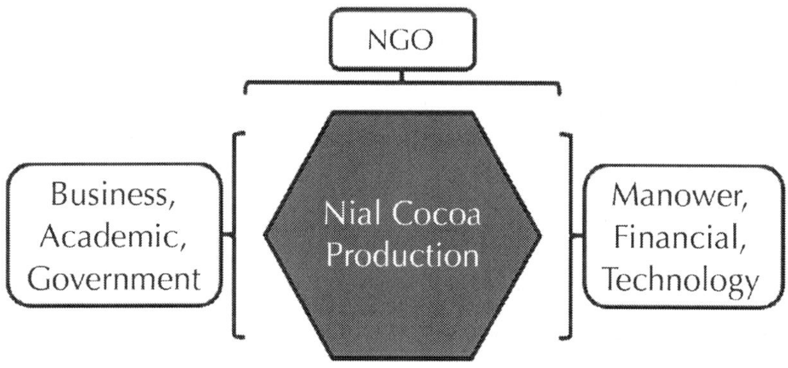

Figure 1: Theoretical Framework

METHODOLOGY

The paper presents a systematic approach which seeks to illustrate the links between parties in a community empowerment project, especially the convergence of business and university which is boosted by government, based on Triple-Helix model. The first section explains about the background what this study address. Then, the second section reviews literatures of each idea. The result is the theoretical framework which is used as the guidance in reaching the main purpose of this study. The following section are used to discuss the convergence between three helix which build the opportunities of each connection based on the theoretical framework, so it will be analyzed carefully through the archived information from internal publications to find the most appropriate relationship in the cycle of field condition. Last section will collect the main idea of this study into several conclusions in a simple model. Finally, further researchers can extend or develop new research based on this study. Ideas can be found in that section to give initial insights.

DISCUSSION

First Phase, Early Approaches

Nias grow with it's unique shape of society. Although includes in the North Sumatra province, the society behave in different ways from North Sumatra native people, Batakese. Understanding the implication of culture from the behavior or the words of Nias people can be too difficult for non-Nias people (Brown, 2003). Although the Indonesian government is an authority which is highly acknowledged locally, the origin of Nias people is seen as non-Indonesian, but shared with the wealthy outsiders (Wolff, 2006). The development of Nias means that NGO should deep-enter the Nias society in smart ways.

As the successor of former NGOs in Nias, they –at the first step– follow the result of another NGOs, make an assessment through field survey, and catch another sensitive issues related to it's work. Schültz & Nugroho (2006) were undertaking a village assessment of the locations where NGOs was working in based on the sustainable livelihoods approach. That is, a detailed assessment were undertaken that provided them with village profiles incorporating the following:

- ***Human Capital***: The level of technical knowledge of cocoa cultivation, education, and literacy.
- ***Natural Capital***: Quantitative assessment of cocoa stocks, condition, and available areas for new planting.
- ***Social Capital***: Ability of farmers to work together, coordinate on issues such as disease control as well as the role of women in the cocoa producer households.
- ***Physical Capital***: Available supporting infrastructure such as roads and bridges.
- ***Financial Capital***: Access to credit, presence of savings.
- ***Value-Chain Processes***: Individual village's position in the value chain of cocoa differs. The NGO want to analyze this to help villages to come up with a future Community Action Plan.
- ***Structure and Organizations***: What are the existing working structures or organizations in the village? What has been tried in the past, what has failed and why?

As identified above, cocoa farmers as the main business persons in cocoa industry are severely restricted by their lack of knowledge and technical expertise regarding more beneficial cultivation, harvesting and post harvesting techniques. So, the NGO planned to facilitate a variety of technical trainings for the first helix –cocoa farmers of Nias– through an activity called "Sekolah Lapang" (field school). It will provide the appropriate techniques for farmers related to the on-farm process of cocoa. The "school" also a ToT (Training of Trainers) in order to prepare the society through the knowledge-transfer. So, there is not only community development but build a community empowerment. The autonomy should behave as the product of empowerment, and it will produce the sustainable development among cocoa society of Nias.

The concept of sekolah lapang was to develop the knowledge of farmers directly on the farm. The experts were invited from many institutions which have tight relationship with cocoa industry, such as M.A.R.S., Inc., ICCRI (Indonesian Coffee and Cocoa Research Institute), and from Indonesian Ministry of Agriculture. They transfered their knowledges how to transform the traditional methods of cultivating cocoa to the modern one, without losing the society identity like multi-commodity farming. As the note, farmers in Nias did not focus only on cocoa. They have many commodity focus on single household (Ülrich, 2009).

From field survey (Schültz & Nugroho, 2006), it's ascertained that the average cocoa farmer owns or has access to around one hectare of vacant land that can be planted with new seedlings. So they decided to supply the farmer's trees in the target areas. The supply will be treated as the supplement of the sekolah lapang. This way the farmers will have enough trees to see the benefits of increased production in the medium term and will be able to graft and selectively propagate from these trees using the cultivation skills in which the experts in sekolah lapang are training them. The survey result was combined with the statistics from department of agriculture to estimate the following number of cocoa farmers in the target areas, and then the supply scheme (included in sekolah lapang) was introduced to the farmers:

- Cocoa focal points are identified in each village. These people were sent to ICCRI for training and will be the main focal point for cocoa activities of the village in the future.
- Agreement between cocoa farmers and NGO. The NGO is willing to supply an amount of cocoa seedlings to each village based on the needs assessment. It also supplied polybags and fertiliser, while the communities supply sand and soil for potting as well as an area for germination beds and seedlings. Complete lists of farmers names and the number of seedlings they will each receive were compiled.
- When the facilities are prepared in the village, NGO purchase and supply the seeds.
- Distribution of seedlings to farmer's occured between 1-2 months after they have been put into polybags.
- A sample of the farmers participating in this activity will also participate in the monthly monitoring of household economic status and primary production activity.

Identification of Surrounding Issues

Market & Product information

There is very limited knowledge among Nias farmers of the value chain of the commodity they produce. That is, farmers are mostly unaware of what happens to their product after it is sold at the market or to the local trader. Without any understanding of the value chains of cocoa from production to consumption, farmers are in a poor position to argue for better prices from their local buyers. Furthermore in order to value add their products through better post-harvest treatment; farmers need to understand the chain more completely. For example, to get value added, it would be beneficial for farmers to understand that cocoa needs to be fermented in order to develop the chocolate flavour of cocoa which occurs when proteins are broken down into amino acids.

Another major issue contributing to low "farmer bargaining power" is the lack of knowledge regarding national commodity prices for cocoa. Very few farmers are accessing this information that is available on Radio Republik Indonesia. NGO have found that the majority of farmers are not aware that this information is available. In some of remote work locations where there is no electricity there are also very few radios. The following activities are being planned to tackle this issue:

1. Distribution of value chain posters for cocoa. In the beneficiary villages NGO distributed graphic posters showing farmers the supply-chain from the raw product to the consumer. The distribution of these materials were done by the Field Facilitator Officers who have been trained in value-chain analysis.
2. Radio Republik Indonesia Stickers. In conjunction with RRI, NGO distributed stickers with clear simple information regarding the time and day that market information can be accessed.
3. Distribution of solar-powered/wind-up radios. In villages that don't have electricity, NGO distributed radios that don't rely on battery or grid electricity.

Gender

In March 2006 the UNDP funded and facilitated a workshop titled "Nias Agricultural Development and Consolidation Workshop," to bring all stakeholders in agriculture together to create action plans for agricultural reconstruction and development in Nias. One of the most profound conclusions from the working group discussions on cocoa was that past attempts do carry out skills trainings in Nias had not had the desired impact because of poor consideration of the gender components of cocoa farming (UNDP, 2008). According to the field surveys, more than 50% of cocoa cultivation and harvesting work is done by women, however, according to sources from this workshop as well as department of agriculture, women have never been directly involved in any skills trainings exercises.

As part of the programme, NGO ensure that women were involved in all aspects of the trainings that carried out, including ToT training courses in Nias as well as outside of Nias. Before planning field trainings for cocoa cultivation and harvesting, the women participants were consulted and asked to propose which potential times and venues they would be most comfortable with and which will not interfere with or increase the burden of other income generating activities and domestic chores. Likewise in the distribution of cocoa seedlings women were consulted to ensure that the village preparations for distribution such as preparing potting mix did not impact on or further burden the women in the village.

Afiluo Dryer: The Engineering and the Acceptance

For most economics practicioners, economic development is often interpreted as an effort to increase the production of goods and services by the society (Nugroho & Hidayat, 2010). Rostow (1960) stated that the increase in output production structure is characterized by the transformation from pure-traditional agriculture to industry. Technology, is always needed to bring multiplier effect for the transformation process. But in Nias, technology should not change, whether just an improvement, the basic behavior of the Nias farmers. Based on the works of Ülrich (2009) and Amin (2009), the NGO then came up with the idea to develop a drying equipment to help farmers get faster drying process of cocoa bean. The main idea was shortening the drying time of cocoa bean. Thus, there should be designed a dryer which could be easily adapted by the farmers, even, with low-level capability or knowledge. The most suited type of technology was the appropriate technology, a technology that accomodate technical aspect and economical aspect, and also include socio-culture aspect –of where the product will be applied– into the engineering process. Appropriate technology was tightly related with applied science subject, so the NGO invited two researcher from two different vocational education institutions to bring the idea into reality. They were the second helix.

After the preliminary and prototyping process of designing the dryer, it was researched and developed in the collaboration with the farmers. In order to get good acceptability level from the society, the dryer was named "Afiluo." The word *afi* means "wings," and *luo* means "sun," so the name afiluo means "a dryer with sun as the power source and wings as the heat collector" (Sianipar, 2010a). Then, after the development process together with the farmers, its appeared that the acceptance of the dryer was fly so high because farmers very excited with the Afiluo and a lot of people volunteered themself to make the dryer. It caused by the facts that the Afiluo could help them to solve their common root problems (Sianipar & Zebua, 2011; Swisscontact, 2011), such as:

- Water Content: There were evidences that the Afiluo could produce higher temperature than traditional method. The temperature was in twice level compared to the temperature outside the dryer. Through higher temperature, cocoa beans were dried into 7%-8% water content, same with SNI (Indonesian National Standard) level, so the farmers could avoid price cutting punishment from the buyers.
- Drying Time: The trial phase of Afiluo shown that the cocoa bean could be dried faster, until 2 days shorter than using traditional dying method, so the post-harvest process could be done in shorter time. Quality increased but did not spend much time.
- Panic Effect: Harvest season in Nias occured in rainy season. The farmers will trapped in panic effect when they use traditional drying method, because they need to recollect the bean when

the rain was come. With many farming products, the farmers could not have enough time to keep the bean safe in all time.

- Bean Color, Taste, And Aroma: They were happened since the fermentation process, but they became perfect because the use of Afiluo. The balance in heat distribution in the bean's body made the taste and aroma spread evenly. Good heat transmissions from the sun to the bean bring beautiful bean's color.
- Animal Disturbance: Using Afiluo, the drying process will be safe from animal disturbance such as dogs, chicken, or pigs, so the waste percentage will decrease, and increase the selling price of dry cocoa bean.
- Manufacturing Cost: Usually, the farmers seems affraid with the high cost of "farm machine." Afiluo was made with appropriate technology approach, so it was easy to build with low cost. Ninety-five percent was made from local materials of Nias archipelago, and 85% was constructed from wood (Sianipar, 2010a).
- Usability: The society will meet a simple equipment. There were no greasing instruction, tank for containing the fuel, even, no electrical instalation. The use of Afiluo would not change existing behaviors.

The Last Helix: Nias Government

From three factors on theoretical framework, the increase of manpower quality and technology were applied as explained before. So, the last factor was the financial investment from the government. The increase of input modal was received from Nias government investation through the PT Bank Sumut as a govermental insitution (BUMN) in North Sumatra province which engaged in financial investation (hereinafter will be referred to as only "BUMN"). In order to build the investation guidance, the NGO and the BUMN develop the purpose and target of the investment process (Swisscontact, 2011). The main points were:

- Fostering microentrepreneurs –farmers– who owned business feasibility but not bankable yet.
- As the critical issue in Nias, the main target was woman from pre-prosperous family or with low-level income who owned micro-business in the cocoa industry and trading. As the note,

that they will receive the investation if their business have good prospect in the future.

The BUMN was develop the investment through credit scheme. There were several important points as the basic provision list, but the most important idea was all of credit members should prove themself as good members (QM-Quality Members) at the time they receive each investation phase. The money should be distributed to another two inputs of production: the knowledge improvement of manpower, and the application as well as replication of dryer as the technology investation.

In other hand, the Nias government through Department of Industry, Trading, Energy, & Natural Resources (Disperindag ESDM) also help to boost the cocoa bean quality through an insitutional movement in the standard application of fermentation in post-harvest cocoa processing (Swisscontact, 2011). The movement had purpose to sensitize the public that good fermentation before the drying process is very important to get high cocoa quality, to reach the export quality level, then receive high selling price.

CONCLUSIONS

The development of triple-helix in Nias cocoa industry was revolutionizing the way of collaboration between stakeholders. The first helix was the the farmers. The farmers behave as the business institution (B) of cocoa production and trading, because they were the main party among the cocoa business. The improvement of farmers' knowledge was about the techniques in cultivating, harvest, and post-harvest processing of cocoa processing. The NGO behave as the facilitator of the knowledge transfer and development through traning for farmers as well as training of trainers (ToT) as the future focal points.

The second helix was the researchers as the representation of university (U). The researchers came to Nias to apply their scientific knowledge into the application of appropriate technology. Their knowledge produced the Afiluo dryer as an equipment for the cocoa farmers to shorten the drying time as well as keep the good cocoa quality after fermentation. The convergence of business and technology in the application of dryer could have such an effect on financial performance, so the knowledge of researcher was kept safe by the

NGO through the scheme that the capitalization of knowledge –the dryer– must be applied only by the farmers directly, not by another commercial institution.

In other hand, the goverment (G) join the triangle as the third helix. Capitalization of knowledge needs financial investment and supporting policy from the government. The Nias government through PT Bank Sumut (BUMN) and Department of Industry, Trading, Energy, & Natural Resources (Disperindag ESDM) made their movements related to the financial investment and also a program in the application of good fermentation and drying process.

In business (B) helix came up Bn, "n" means the business institution was consist of many farmers. Bn interacted with another Bn and also with the NGO in their circle. University helix (U) consist of two researchers from two different institution (U1 and U2). Their collaboration was met with the NGO as the party who invited them to capitalized their knowledges. The last helix, Government (G) made their movements through two insitutions (G1 and G2). Their join linkage with the cocoa production was connected through the NGO.

Finally, the NGO would be the axis of those three helix. The collaboration between farmers as business institutions (B/Bn), the university through the researchers (U1 and U2), and government through G1 and G2 became the cycle in which coordinated by the NGO. As a cycle, the collaboration continuously develop themself to reach better and better future. In the end of the project, December 2011, the NGO can leave the Nias society with the the belief that they not only build Nias, but also empowered the people there so they can achieve sustainable prosperity.

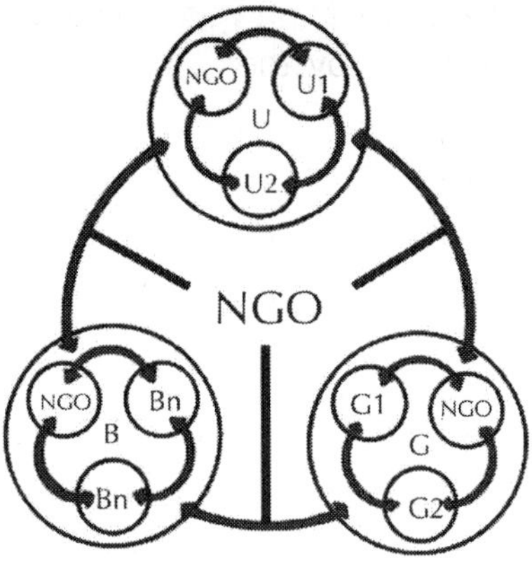

Figure 2: NGO: The Triple-Helix Axis

FURTHER RESEARCH

This study should be continued through many further researches. The new thought about the application of Triple Helix model will bring the widening of Triple-Helix approach in the community empowerment study. At the future, researcher maybe conducts new research about contra-productive effect of Triple-Helix approach in another project. In other hand, other NGOs may start their projects in community empowerment based on Triple-Helix approach but on another project purposes or objects. Above all, another genius research project may try to do a generalization of Triple-Helix model in community empowerment as well as community development projects among developing countries.

REFERENCES

1. Adam and Negara, 2010 Adam, L. & Negara, S.D. (2010). ASEAN-China Free Trade Agreement: Tantangan dan Peluang bagi Indonesia. *Masyarakat Indonesia*, XXXVI(2), 1-24. in press.
2. Amin., 2009 Amin (2009). *Teknik Pengeringan Kakao*. Internal publication. Nias: Swisscontact-NISA. in press.
3. Asia Pulse, 2011 Asia Pulse (2011). Indonesia to become Worlds

Largest Cocoa Producer. *Asia Pulse,* February 03. Rhodes: Asia Pulse Pty Ltd. in press.

4. Badan Pusat Statistik (BPS), 2009 Badan Pusat Statistik (BPS) (2009). Nias dalam Angka: Annual Report 2008. Nias: Badan Pusat Statistik. in press.
5. de Blij and Müller, 2010 de Blij, H.J. & Müller, P.O. (2010). *Geography: Realms, Regions, and Concepts.* New Jersey: Wiley & Sons, Inc. in press.
6. Brown, 2003 Brown, L. (2003). Li Niha - The Nature of Culture, Tradition, and History of Nias People.*Ya'ahowu News*, November 10. in press.
7. Clay, 2004 Clay, J. (2004). *World Agriculture and the Environment.* Washington DC: Island Press. in press.
8. Dalle, 2003 Dalle, J.M. (2003). *Open Source Technology Transfer.* Paper presented at the 3rd EPIP Conference, Maastricht, NLD, 22-23 November. in press.
9. Donald, 2004 Donald, P.F. (2004) Biodiversity Impacts of Some Agricultural Commodity Production Systems. *Conservation Biology*, 18, 17-37. in press.
10. Etzkowitz, 2008a Etzkowitz, H. (2008). The Triple Helix: University–Industry–Government Innovation in Action. London: Routledge. in press.
11. Etzkowitz, 2008b Etzkowitz, H. & Viale, Riccardo (2008). Introduction: Anti-cyclic Triple Helix, in Henry Etzkowitz & Riccardo Viale (eds.)., *The Capitalization of Knowledge: A Triple Helix of University–Industry–Government*, pp. 1-27. in press.
12. Firdausy, 2010 Firdausy, C.M. (2010). Konsep dan Kebijakan Good Governance: Suatu Tinjauan Ekonomi. *Jrnl. Eko dan Pembangunan*, XVII(2), 13-21. in press.
13. Fitch Ratings, 2011 Fitch Ratings (2011). Indonesia: Full Rating Report. New York: Fitch, Inc., Fitch Ratings Ltd, December 2011. in press.
14. Gambardella and Hall, 2010 Gambardella, A. & Hall, B.H. (2010). Proprietary versus Public Domain Licensing of Software and Research Products, in H. Etzkowitz & R. Viale (eds.)., *The Capitalization of Knowledge: A Triple Helix of University–Industry–Government*, pp. 167-198. in press.

15. Hutabarat et al., 2007 Hutabarat, et al. (2007). Analisis Kesepakatan Perdagangan Bebas Indonesia-China dan Kerjasama AFTA serta Dampaknya terhadap Perdagangan Komoditas Pertanian Indonesia. Bogor, Indonesia: R&D Agency, Ministry of Agriculture, Republic of Indonesia. in press.
16. ICCO., 2007 ICCO (2007). International Cocoa Organization: Annual Report 2007, pp. 43. in press.
17. Maurer, 2002 Maurer, S.M. (2002), *Promoting and Disseminating Knowledge: the Public/Private Interface*. Paper presented to the U.S. National Research Council's Symposium on the Role of Scientific and Technical Data and Information in the Public Domain, Washington, D.C., 5-6 Sept. in press.
18. Moser et al., 2010 Moser, *et al.* (2010). Response of Cocoa Trees (Theobroma Cacao L.) to a 13-month Desiccation Period in Sulawesi, Indonesia. *Agroforestry Systems*, 79(2), 171-187. in press.
19. Ng, 2011 Ng, R. (2011). A speed date with chocolate. *Chemical Engineering Progress*, 107(7), 17. in press.
20. Nugroho and Hidayat, 2010 Nugroho, A.E. & Hidayat, A.S. (2010). Good Governance dalam Pengelolaan Investasi di Provinsi Sumatera Selatan. *Jrnl Eko dan Pembangunan*, XVII(2), 45-61. in press.
21. Rostow, 1960 Rostow, W.W. (1960). *The Stages of Economic Growth: A Non-Communist Manifesto*, chapter 2, "The Five Stages of Growth: A Summary," pp. 4-16. Cambridge: Cambridge University Press. in press.
22. Schültz and Nugroho, 2006 Schültz, M. & Nugroho, P.A. (2006). *Oxfam Nias Strategy to Help Poor Cocoa and Rubber Farmers*. Unpublished assessment. Nias: Oxfam. in press.
23. Sianipar, 2010a Sianipar, C.P.M. (2010a). *Manual Preparation and Use of "AfiLuo" Cocoa Beans Drying Equipment*. Internal. Nias: Swisscontact-NISA. in press.
24. Sianipar, 2010b Sianipar, C.P.M. (2010b). *The Engineering of Cocoa Bean Dryer in the Nias Island*. Undergraduate Thesis. Bandung, Indonesia: Bandung Polytechnic for Manufacturing (POLMAN Bandung). in press.
25. Sianipar and Zebua, 2011 Sianipar, C.P.M. & Zebua, D. (2011).

Technical Recommendation and Product Documentation of "AfiLuo" Cocoa Beans Drying Equipment. Internal. Nias: Swisscontact-NISA. in press.

26. Swisscontact, 2010 Swisscontact (2010). Swisscontact: Annual Report 2009. Internal publication. Jakarta: Swisscontact-Regional Office SE Asia. in press.
27. Swisscontact, 2011 Swisscontact (2011). *Afiluo Dryer: Alat Pengering Biji Kakao*. Nias: Forum Kakao Kepulauan Nias, Disperindag ESDM Kabupaten Nias, & PT Bank Sumut cabang Gunungsitoli. in press.
28. Thaler and Sunstein, 2008 Thaler, R.H. & Sunstein, C.R. (2008). *Nudge: Improving Decisions About Health, Wealth, and Happiness*. CT: Yale University Press. in press.
29. Ülrich, 2009 Ülrich, T. (2009). *Cocoa Post Harvest Technology Assessment*. Unpublished assessment. Nias: Swisscontact-NISA. in press.
30. UNDP, 2008 UNDP (2008). Indonesia: Crisis Prevention and Recovery Update. Jan 2008. Jakarta, Indonesia: United Nations Development Program. in press.
31. Viale, 2010 Viale, R. (2010). Knowledge-driven Capitalization of Knowledge, in Henry Etzkowitz & Riccardo Viale (eds.)., *The Capitalization of Knowledge: A Triple Helix of University–Industry–Government*, pp. 31-73. in press.
32. Wall Street Journal, 2011 Wall Street Journal (2011). Indonesia, Ghana may Fill Cocoa Void. Wall Street Journal, 2011, April 06, pp. n/a. in press.
33. Wolff, 2006 Wolff, B. (2006). The Local Organisation and Understanding of Tourism on Nias Island. Denmark: Museum of Denmark. in press.
34. Wood, 2010 Wood, L. (2010). Research and Markets: Indonesia Agribusiness Report Q1 2011 - Cocoa Production Growth to 2014/15: 55% and Coffee Production Growth to 2014/15: 10%. Business Wire, 2010, November 30. in press.

Citations

CHAPTER 1

Simplice Tagro Guehi, Michael Dingkuhn, Emile Cros, Gérard Fourny, Robert, Ratomahenina, Guy Moulinand Anne Clement Vidal, Impact of Cocoa Processing Technologies in Free Fatty Acids Formation in Stored Raw Cocoa Beans, ISSN 1991-637X.

CHAPTER 2

Oscar Orlando Ortiz-R; Raquel Amanda Villamizar Gallardo; Joshua Mauricio Rangel, Applying Life Cycle Management of Colombian Cocoa Production, http://dx.doi.org/10.1590/S0101-20612014005000006.

CHAPTER 3

Ndife Joel, Bolaji Pius, Atoyebi Deborah and Umezuruike Chris, Production and quality evaluation of cocoa products (plain cocoa powder and chocolate), doi:10.5251/ajfn.2013.3.1.31.38

CHAPTER 4

J. Amang A Mbang, P. Mounjouenpou, E. Nossi, D. Achukwi, S. Bassanaga, and N. Woin, Adoption of new Technologies of Traditional Transformation of Products Made from Cocoa (Theobroma cacao L.) in Cameroon, ISSN: 2231-5381.

CHAPTER 5

Pauline Mounjouenpou, Dominique Gueule, Suzanne Abeline Maboune Tetmoun, Bernard Guyot, Angélique Fontana-Tachon, and Joseph Pierre Guiraud, Incidence of Pod Integrity on the Fungal Microflora and Ochratoxin-A Production in Cocoa, ISSN 2157-6076.

CHAPTER 6

Pauline Mounjouenpou, Joseph Amang A Mbang, Bernard Guyot , Joseph-Pierre Guiraud, Traditional Procedures of Cocoa Processing and Occurrence of Ochratoxin - A in the Derived Products, ISSN : 0975-7384.

CHAPTER 7

Mohammed, B. T., Achem, B. A., and Abdulquadri, A. F., Factors Influencing Adoption of Agricultural Processing Technologies Developed by National Centre for Agricultural Mechanization (NCAM) in Ifelodun Local Government Area, Ilorin Kwara State, ISSN (Online): 2319-7064.

CHAPTER 8

A. Bart-Plange, A. Addo, S. K. Amponsah & J. Ampah, Effect of Moisture, Bulk Density and Temperature on Thermal Conductivity of Ground Cocoa Beans and Ground Sheanut Kernels, ISSN: 2249-4626.

CHAPTER 9

Effendy, Nuhfil Hanani, Budi Setiawan & A. Wahib Muhaimin, Effect characteristics of farmers on the level of technology Adoption side-grafting in cocoa farming at sigi regency-Indonesia, ISSN 1916-9752

CHAPTER 10

David Koffi Akaki, Eric Fattoh Elleingand, Sadat AW, and Coulibaly Kelegoun, Color Characteristics of Cocoa Butter Made From the Processing of the Ivorian Cocoa Beans, ISSN: 0975-5462.

CHAPTER 11

Louis Koko, Teractiv Cacao as a new fertilizer based Reactive Phosphate Rock for cocoa productivity in Côte d'Ivoire: A participatory approach to update fertilization recommendation, doi: 10.1016/j.proeng.2014.09.027

CHAPTER 12

Elizabeth Chin, Kenneth B Miller, Mark J Payne, W Jeffery Hurst, and David A Stuart, Comparison of antioxidant activity and flavanol content of cacao beans processed by modern and traditional Mesoamerican methods, doi:10.1186/2050-7445-1-9.